RICK PETERS

FRAMING BASICS

Main Street
A division of Sterling Publishing Co., Inc.
New York

Acknowledgements

Butterick Media Production Staff

Photography: Christopher Vendetta
Design: Triad Design Group, Ltd.
Illustrations: Triad Design Group, Ltd.
Copy Editor: Barbara McIntosh Webb
Page Layout: David Joinnides

Indexer: Nan Badgett
Project Director: David Joinnides
President: Art Joinnides
Proofreader: Nicole Pressly

Special thanks to Peter Lamia at MarinoWare for his help with the WareWall metal framing illustrations, Bill Georges at Simpson Strong-Tie for his help with the metal framing connector illustrations, and Erik Wilson with the Western Wood Products Association for his help with the joist and rafter span charts. Also, thanks to the production staff at Butterick Media for their continuing support. And finally, a heartfelt thanks to my constant inspiration: Cheryl, Lynne, Will, and Beth. **R. P.**

Every effort has been made to ensure that all the information in this book is accurate. However, due to differing conditions, tools, and individual skill, the publisher cannot be responsible for any injuries, losses, or other damages which may result from the use of information in this book.

Library of Congress Cataloging-in-Publication Data Available

Published by Sterling Publishing Company, Inc.
387 Park Avenue South, New York, N.Y. 10016
© 2000 Butterick Company, Inc., Rick Peters
Distributed in Canada by Sterling Publishing
c/o Canadian Manda Group
One Atlantic Avenue, Suite 105
Toronto, Ontario, Canada M6K 3E7
Distributed in Great Britain by Chrysalis Books
64 Brewery Road, London N7 9NT, England
Distributed in Australia by Capricorn Link
(Australia) Pty. Ltd.
P.O. Box 704, Windsor, NSW 2756 Australia

10 9 8 7 6 5 4 3 2 1

Printed in China
All rights reserved

Main Street ISBN 1-4027-1088-7

Contents

Introduction

Of all the home improvement challenges that the average homeowner faces, moving, removing, or installing a wall is perceived as one of the most daunting. That's because of "hidden" surprises that everyone believes are lurking behind an existing wall, just waiting to jump out and throw the proverbial monkey wrench into the job.

I'll be honest with you: Most walls are not empty—ductwork, electrical, plumbing, and gas lines are commonly found within. But finding something inside doesn't have to be a surprise. With a little bit of detective work and a solid understanding of how a house is built, you can plan for these—and easily overcome any complications that they might create.

Altering a wall, particularly a non-load-bearing or partition wall, is fairly straightforward. Walls that do support the weight of the structure, or load-bearing walls, are another matter entirely. This is where building codes, permits, and inspectors come into play. It's absolutely imperative that you contact your local building inspector if you're planning on modifying a load-bearing wall. I've worked with a lot of these folks over the years, and I've found them to be a friendly and helpful group of people.

In this book, I'll start by taking you through the different framing systems in Chapter 1. I'll go over the three most common types of house framing—platform, balloon, and post-and-beam—and help you identify which type your home has. Then, so that you'll have a solid grasp of framing terminology and construction methods, I'll cover the various framing systems—foundations and floors, walls and partitions, and roofs. You'll also find information on how to read blueprints, including abbreviations, symbols, and plan views. Finally, I'll discuss codes, permits, and inspections, all of which are designed to protect both you and your family.

Chapter 2 is all about the tools and materials you'll use for framing. I'll decribe the tools you'll need for most framing jobs—everything from layout and measuring to demolition. In terms of materials, solid lumber is not the only choice for framing anymore. Metal framing and engineered panels are stronger and often less expensive than solid wood. Regardless of what material you choose, the members will need to be fastened together: Nails, screws, anchors, and hangers are all discussed in detail in the final section.

In Chapter 3, I'll describe techniques used for framing. I'll begin with layout and measurement (tape measures, squares, levels, and plumb bob)—everything you need to know to lay out and define a project successfully. Then on to tool techniques: nailing with a hammer or an air-powered nailer, and using hand-, circular, reciprocating, and saber saws. There's also a section on working with metal framing and tips on straightening lumber and fine-tuning joints.

Chapter 4 looks at framing joints in detail, including how to frame corners and what to do when walls intersect. Next, there's information on making headers and framing rough openings for windows and doors, installing plating, even adding floor sheathing. Get out your sledge-hammer and roll up your sleeves for Chapter 5— Demolition. There's more to this job than just Popeye-like strength. You have to know how something is built in order to take it apart. I'll share numerous tips and techniques to make this messy task go as smooth as possible.

When the dust clears, you'll often want to add a new wall or wall section. Chapter 6 will take you through the necessary steps one at a time. In addition to how to add a non-load-bearing wall, there's information on how to install a knee wall in the attic and how to finish off basement walls.

Chapter 7, Adding Built-Ins, is for those homes that suffer from lack or storage space (that would be everyone's, yes?). One of the slickest ways to add much-needed space is to take advantage of unused space within or between your existing walls. I'll show you how to locate these spaces and then how to add a built-in—everything from a built-in window seat or a wall storage unit to a new closet.

All in all, I hope that this book encourages you to take on a framing job that you've hesitated to tackle in the past. Altering a wall, adding a door or window, or installing a built-in window seat can have a profound impact on your living spaces. I hope that *Framing Basics* helps you with your home improvement adventures.

Rick Peters
Fall 2000

Chapter 1
Framing Systems

In order to take on a variety of home improvement projects, including basic framing, you'll need to have a solid understanding of framing systems—that is, how a house is put together from the ground up. There are three main framing systems in your home: foundation and floors, walls and partitions, and ceilings and roofs. Each of these systems has its own terminology and methods of construction. For example, if you're planning on modifying one of the walls in your house, you'll need to know whether it has platform or balloon framing.

In this chapter, I'll start by going over the three main methods used to frame a house: platform framing, balloon framing, and post-and-beam. Platform framing is the most common type of framing used in modern residential construction (*see the opposite page*). It uses standard dimension lumber, goes up quickly, and, by the nature of its design, helps prevent fire from spreading within a structure. This "fire blocking" ability is why platform framing replaced balloon framing (*page 8*). Since wall studs in balloon framing run nonstop from foundation to roof, fire could easily spread from one floor to another. But

balloon framing was a big step forward for the construction industry over post-and-beam construction (*page 9*), where heavy timbers were used and great skill was required to join these pieces together.

Next, I'll take you through the common terms and methods of construction used to frame the various parts of a house. Foundations and floors are on pages 10–12, walls on pages 13–15, and roofs and roof types on pages 16–19.

Every new construction job begins with a set of detailed drawings or blueprints that describe the work to be done. I'll go over the abbreviations, symbols, and plan views that will help you visualize a home improvement task (pages 20–23). Finally, there's an important section that covers codes, permits, and inspections, on pages 24–25: everything from sample permits to the type of work that absolutely must meet code requirements and be inspected by local authorities.

Platform Framing

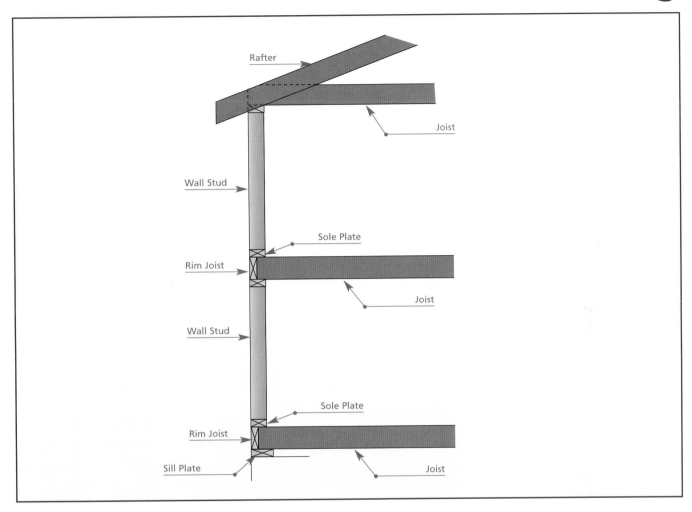

Platform framing is the most common construction method used today to build homes and other structures. A platform-framed structure is built one story at a time; each story is built upon a platform that consists of joists and a subfloor; *see the drawing above.* When one story is completed, the next platform is built, and then construction on the second story can begin.

The separate platform of each story—technically the sole plate and top plate of each layer—eliminates the primary disadvantage of balloon framing: the lack of fire-stops (*see page 8 for more on this*). The top and sole plates also provide convenient nailing surfaces for the installation of drywall and other wall coverings.

Platform framing uses less wood than other methods, maximizes interior space, eliminates bridging, allows single-layer floors and simpler corner posts—it's no wonder this method is the most popular.

Balloon Framing

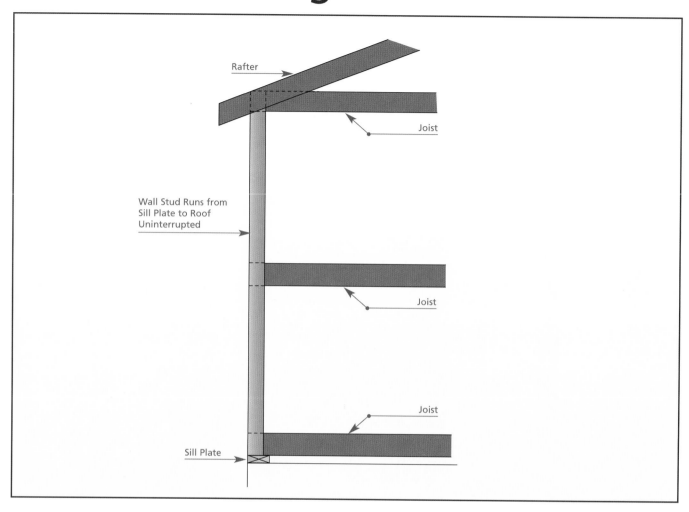

Rafter

Joist

Wall Stud Runs from
Sill Plate to Roof
Uninterrupted

Joist

Joist

Sill Plate

The identifying feature of a balloon-frame structure is that the wall studs run unbroken from sill to top plate, no matter how many stories the structure has; *see the drawing above.* This creates a membrane-like or balloon-like frame. Although many older homes were constructed this way, balloon-framed houses have been superseded by the newer platform framing method (*see page 7*).

In its day, balloon framing offered a number of advantages over the post-and-beam structure: It took advantage of manmade nails, which were a lot cheaper than hand-pegged joinery; it used inexpensive standardized lumber; and it created stable exteriors that were easy to cover with stucco and other materials.

But balloon framing had some problems. First, blocking needed to be added in the vertical wall cavities to prevent fires from spreading from floor to floor. Also, insulating this type of wall was difficult, as it usually opened into the basement at the sill.

Post-and-Beam Framing

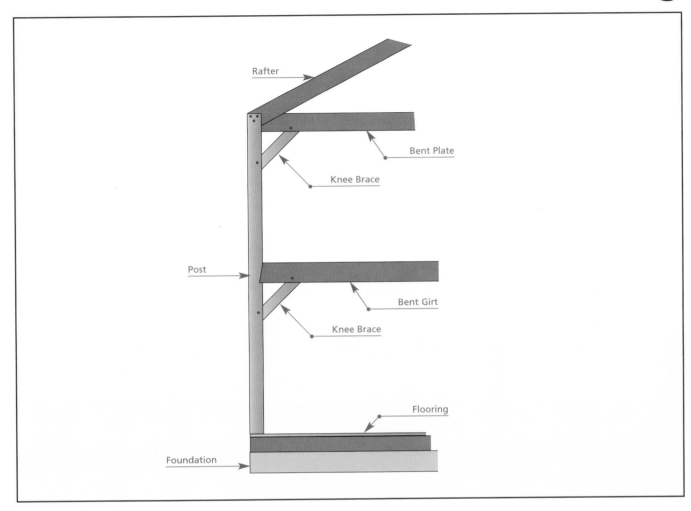

Rafter

Bent Plate

Knee Brace

Post

Bent Girt

Knee Brace

Flooring

Foundation

Post-and-beam construction is easily identified by its use of large, widely spaced load-carrying timbers; *see the drawing above.* Post-and-beam construction is also referred to as post-and-girt or post-and-lintel and was developed in Europe well before it became popular in the United States. The stability and strength of this method of construction can be seen in century-old houses that are still standing—and will remain standing for years to come.

In a post-and-beam structure, the vertical loads are supported by posts at the corners and intersections of load-bearing walls. The weight of the floor and roof, by way of the rafters and joists, is collected by the plates and girts and is then transferred to the posts. (Unlike those in platform and balloon-framed structures, the wall studs in a post-and-beam structure carry no weight and are therefore not load-bearing.)

The primary method used to join these heavy timbers is the mortise-and-tenon joint. Because of this, post-and-beam construction requires specialized builders or "joiners" to handle the complex joinery.

Foundations and Floors

To best support the load or weight of the structure, typically its foundation rests on concrete pads or footings. The depth and width of these footings will depend on your location; see your local building inspector for code specifications in your area. In locations that are frost-free, the foundation can be a slab that will also double as the subfloor. Footings are still required for a slab around the perimeter and under any interior load-bearing wall.

Floor framing is attached directly to the foundation, beginning with the mudsill. The mudsill lies face-up on the foundation and is usually secured to it with metal anchor bolts that are spaced evenly apart. Band or rim joists are then installed on top of the mudsill and provide a nailing platform for joists and the subfloor. Although floor framing is fairly straightforward and simple, it must be installed precisely since any imperfections will continue into the floors, walls, and roof of the structure.

Common flooring terms

Term	Definition
Beam	A steel or wood member that's installed horizontally to support part of a structure's load
Bridging	Steel braces or wood blocks that are installed in an X-pattern between floor joists to stabilize the joists
Column	A metal or wood vertical member designed to support part of a structure's load
Footing	Typically a poured concrete base on which the foundation of a structure is built
Girder	A horizontal steel or wood member used to support part of a structure's load
Joist	Framing lumber that's installed horizontally on edge, to which subfloors are attached
Pier	A round or square concrete base that supports columns, posts, girders, or joists
Rim Joists	The joists that define the outside edges of a platform, typically nailed to the ends of floor joists
Subfloor	Plywood or oriented-strand board that's structurally rated for flooring and is installed over the joists
Underlayment	A thin layer of stable, often water-resistant plywood or other material that's installed on top of the subfloor to create a smooth base for the floor covering

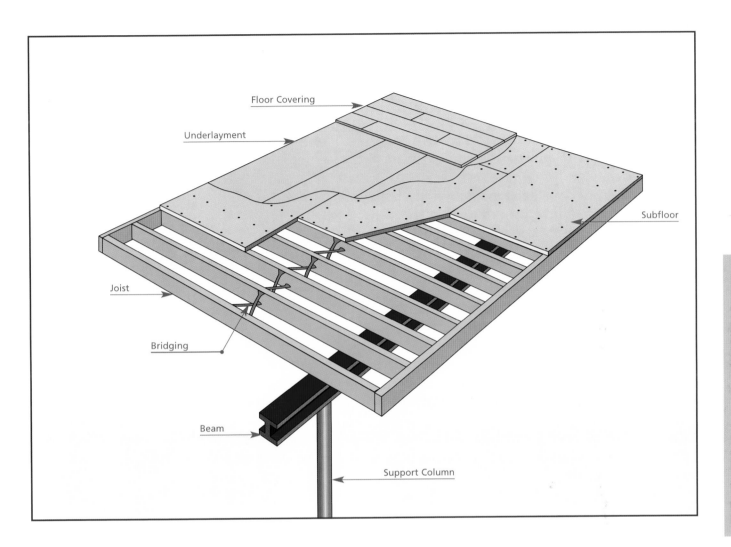

Floor Covering

Underlayment

Subfloor

Joist

Bridging

Beam

Support Column

Anatomy of a Floor

Regardless of the material chosen as the top layer, the underlying structure of most floors is similar. The most common type in residential construction is the framed floor. On a ground-level framed floor, the flooring rests on joists that sit on sills along the foundation and is often supported at a midpoint by a girder.

An elevated framed floor like the one *shown above* is supported by beams that run perpendicular to the joists, where the weight of the floor is borne by support columns. In most cases, the joists are tied together with bridging for

extra stability and to keep them moving from side to side.

The joists are covered with some form of sub-flooring, typically tongue-and-groove plywood, particleboard, or OSB (oriented-strand board). Depending on the type of flooring used, the subfloor may be covered with an additional layer of underlayment (*see page 12*), such as cement board. The top layer of flooring is installed on top of the underlayment or subfloor and usually rests on some type of cushioning layer such as roofing felt.

Post and Girder When the span that floor joists need to go across is too great to go unsupported, often a post-and-girder system is installed to provide the necessary support (*see the photo at right*). The post is the vertical member and may be solid wood (*as shown*), built-up wood, or concrete-filled steel columns (referred to as Lally tubes). The standard wood post rests on a concrete pier or metal post anchor embedded in a footing. A horizontal girder rests on top of the post to support the joists. It may be solid wood, built-up wood, or a steel beam.

Floor Joists Floor joists typically span the distance from foundation wall to foundation wall. They can be 2×6, 2×8, 2×10, or 2×12 solid wood or engineered joists (*see page 34*). Floor joists are evenly spaced, and their spacing and sizing are determined by local codes; *see the chart on page 25 for recommended spans.* To add stability, steel or wood bridging is often installed between joists at regular intervals. In older homes, joists are usually nailed directly to the rim joists. In modern homes, the joists typically rest in joist hangers (*see page 40*).

Underlayment Although platform framing does allow for single-layer floors, an additional layer called the underlayment is installed on top of the subfloor to create a perfectly flat reference surface for a new floor. To prevent cracks in the new flooring, it's important that the underlayment be installed so that its seams do not match up with the seams in the subfloor. The underlayment is cut as necessary to prevent this from happening (*see the drawing at right*).

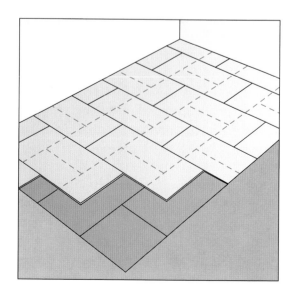

Walls

Framing a wall is fairly simple—typically wall studs are spaced 16" on center (OC) and tied to a single sole plate attached to the subfloor and one or two top plates. In the past, often these framing members were all 2×4 stock. This helped standardize many of the wall framing practices. But today, with greater emphasis being placed on energy efficiency, more and more homes are being built with 2×6 framing. This has a couple of advantages. First, the deeper wall cavity allows for thicker, more energy efficient insulation to be installed. Second, since the studs are beefier, they can be spaced at 24" intervals instead of every 16".

Exterior walls are heavily insulated and have a vapor barrier installed on the warm side of the wall (typically a thin sheet of continuous plastic) to prevent moisture from entering the house. The cold side of the wall can be further insulated with rigid foam board over which exterior siding is then installed. Interior walls are often left uninsulated and are covered directly with drywall or another wall covering.

Common wall terms

Term	Definition
Blocking	Horizontal blocks that are inserted between studs every 10 vertical feet to prevent the spread of fire in a home
Cripple Studs	Short vertical studs installed between a header and a top plate or between the bottom of a rough sill and the sole plate
Double Top Plate	A double layer of 2-by material running horizontally on top of and nailed to the wall studs
Header	A horizontal framing member that runs above rough openings to take on the load that would have been carried by the wall studs; may be solid wood, built up from 2-by material, or an enginnered beam such as MicroLam or GlueLam
Jack Stud	A stud that runs between the sole plate and the bottom of the header; also referred to as a trimmer stud
King Stud	The wall stud to which the jack stud is attached to create a rough opening for a window or door
Rough Sill	A horizontal framing member that defines the bottom of a window's rough opening
Sheathing	Panel material, typically plywood, that's applied to the exterior of a wall prior to the installation of siding
Sole Plate	A horizontal 2-by framing member that is attached directly to the masonry foundation or flooring; also referred to as a sill plate or mudsill
Stud	A vertical 2-by framing member that extends from the bottom plate to the top plate in a stud wall
Top Plate	A horizontal 2-by framing member that's nailed to the tops of the wall studs

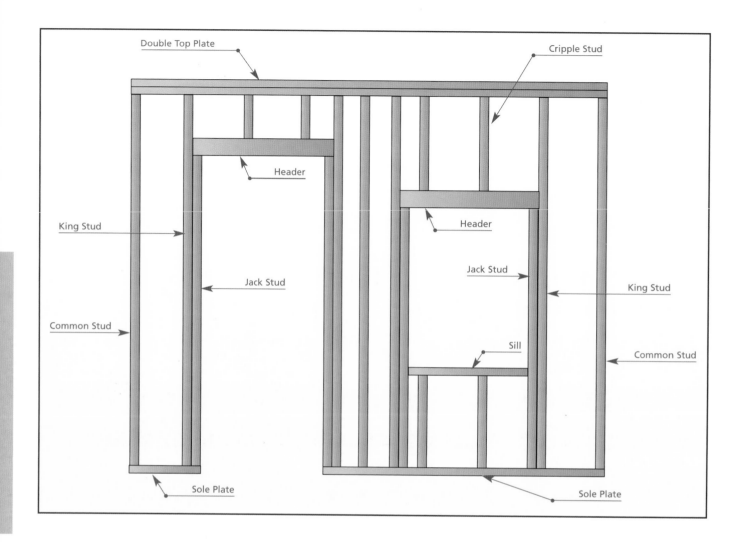

Double Top Plate

Cripple Stud

Header

Header

King Stud

Jack Stud

Jack Stud

King Stud

Common Stud

Sill

Common Stud

Sole Plate

Sole Plate

Anatomy of a Wall

A typical 2-by wall consists of vertical wall studs that run between the sole plate attached to the subfloor and the top plate or double top plate; *see the drawing above.*

Whenever an opening is made in the wall for a window or door, a horizontal framing member called a header is installed to assume the load of the wall studs that were removed. The header is supported by jack studs (also referred to as trimmer studs) that are attached to full-length wall studs known as king studs.

The shorter studs that run between the header and the double top plate or from the underside of the rough sill of a window to the sole plate are called cripple studs.

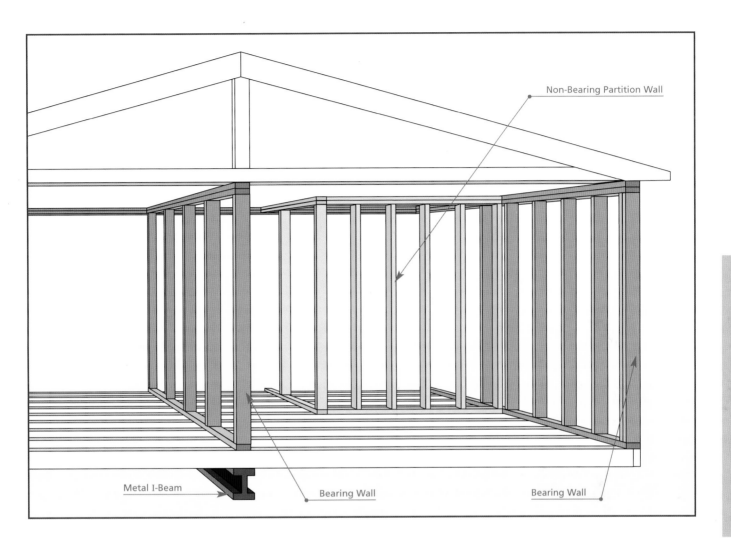

Non-Bearing Partition Wall

Metal I-Beam

Bearing Wall

Bearing Wall

Bearing vs. Non-Load-Bearing Walls

The walls in a structure can be classified into one of two categories: load-bearing and non-load-bearing. A load-bearing wall helps support the weight of a house; a non-load-bearing wall doesn't.

All of the exterior walls that run perpendicular to the floor and ceiling joists in a structure are load-bearing walls because they support joists and rafters either at their ends or at their midspans (*see the dark brown walls in the drawing above*).

Also, any interior wall that's located directly above a girder or interior foundation wall is load-bearing (the center wall in the drawing that sits directly above the steel I-beam).

Non-load-bearing walls, often referred to as partition walls, have relaxed design parameters and code requirements such as wider stud spacing (24" vs. 16" on center) and smaller headers, since they don't support any of the structure's weight (*see the light brown walls in the drawing above*).

Roofs

Framing a roof is not easy: It involves many calculations, rafter tables, and cutting lumber at precise angles. If you need to frame a roof, I'd suggest you use premade trusses (*see page 18*) or hire out this part of the job to a licensed contractor.

Code requirements for roof framing are very strict because there's a lot of engineering involved to make sure the framing can support the roof's weight year-round (including maybe a couple feet of snow). A roof is something you just shouldn't mess around with—get professional advice.

Most roof designs are based on the gable roof (*see page 17*). Once framed, the rafters are

covered with sheathing and the gable ends are enclosed with siding. After the sheathing is in place, roofing felt is stapled on and a covering is added—usually asphalt shingles. To protect the edge of the roof, metal drip edge is often added to funnel water away from the edges and the exterior walls below (*see the photo above*).

Common roofing terms

Term	Definition
Bird's Mouth	A notch cut near the end of a rafter to fit over a cap plate
Chords	Framing members that make up the two sides of the roof and the base of a triangular truss
Collar Tie	A horizontal framing member installed bewteen rafters to add stiffness
Cornice	The part of the roof that overhangs a wall; also called the roof overhang
Dormer	A shedlike structure that projects out from a roof to provide additional attic space
Drip Edge	A bent metal strip that fits over the edge of the roof to direct rain away from the roof edge and underlying walls
Eaves	The part of the roof that projects past the supporting walls
Fascia	A trim piece fastened to the ends of the rafters to form part of the cornice
Flashing	Thin metal used to bridge gaps between the roof and framing or the shingles and framing; also used to line valleys to shed water
Frieze Board	Trim pieces installed directly beneath the rafters to provide a nailing surface for the soffits and corner trim
Hip Rafter	Any rafter that runs at a 45° angle from the end of the ridge to a corner of the structure
Jack Rafter	A short rafter that runs between two rafters or a rafter and a top plate
Overhang	The end of the rafter that projects beyond the building line; typically enclosed by a soffit
Pitch	The rise of the roof over its span
Ridgeboard	The horizontal board that defines the roof's highest point or ridge
Rise	The vertical distance between the supporting wall's cap plate and the point where a line, drawn through the outside edge of the cap plate and parallel to the roof's edge, intersects the centerline of the ridge board
Slope	The rise of the roof over its run expressed as the number of inches of rise per unit of run (typically 12"); 8 in 12 means a roof rises 8" for every 12" of run
Soffit	The board that runs the length of a wall, spanning between the wall and the fascia on the underside of the rafters

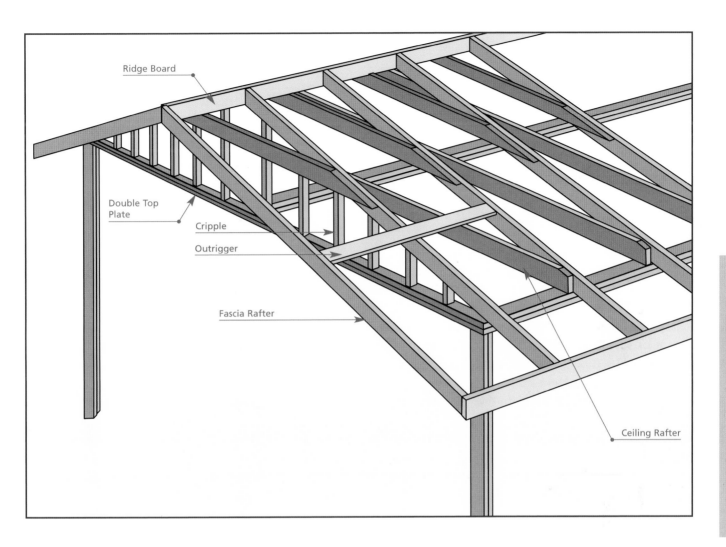

Ridge Board

Double Top Plate

Cripple

Outrigger

Fascia Rafter

Ceiling Rafter

Anatomy of a Roof

Although roof framing is beyond the scope of this book, an understanding of the common parts of a roof is beneficial to anyone tackling a framing job in their home.

The roof *shown here* is a simple gable roof where the two halves of the roof slope in opposite directions. A ridge board runs along the peak of the roof. Common rafters connect to the ridge board and extend down to the double top plates of the walls; they typically tie into the ceiling joists, which prevent the walls from bowing out under the load of the roof.

Gable studs run vertically between the double top plate and the end rafters to form the gable. An overhang is often created by extending the roof past the wall with lookouts that support the barge rafter. The ends of the common rafters are typically covered with trim pieces called the fascia.

Trusses Trusses are designed to support a roof over a wide span (*see the drawing at right*). Since this eliminates the need for load-bearing partitions below, trusses are used extensively in residential construction. In addition to this, the use of trusses greatly simplifies framing, and the roof can be put up quickly. One disadvantage to using roof trusses is that by the nature of their design, much usable attic space is lost.

All trusses consist of three main parts: upper chords that serve as rafters, lower chords that act as ceiling joists, and web members that tie the chords together. The parts of the truss are typically held together with metal or wood gusset plates. The king post truss (*top in drawing*) is a simple truss system that can only span distances less than 25 feet. Fink trusses (*middle in drawing*) are the most common and can span over 40 feet. A scissors truss (*bottom in drawing*) can also span 40 feet and allows for slightly more interior space.

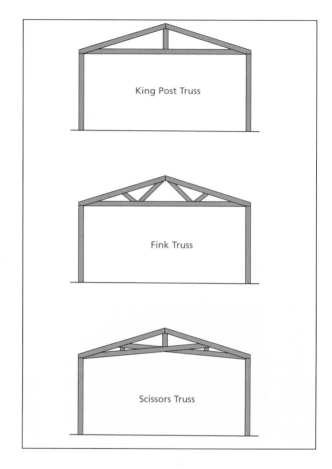

Rafters Without a doubt, the roof is the most complicated part of any framing job for a structure, typically because of the multiple angled cuts required for the rafters that run between the ridge and the top plates (*see the drawing at right*). Adding to the confusion is the specialized names of many of the parts.

Common rafters (*blue in the drawing*) run the full distance from the ridge (*yellow in drawing*) to the double top plates (*tan in drawing*). Hip rafters (*green in drawing*) connect to the ridge at an angle where the two planes of the roof meet. The jack rafters (*red in drawing*) run from the hip rafter to the eave and don't support much of the roof load.

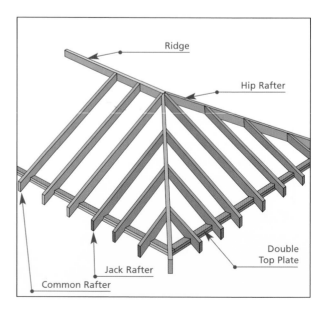

Roof Types

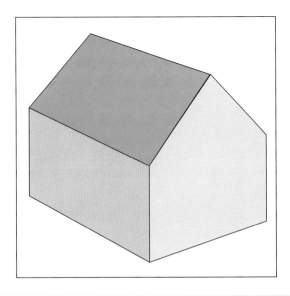

Gable The gable roof *shown in the drawing at left* is the most common roof style used in home construction. Its two sloping surfaces meet at the top and form triangular shapes at each end of the building, called gables. The gable roof is easy to frame and sheds water well. There are a number of variations of this style, the most common being the Tudor peak.

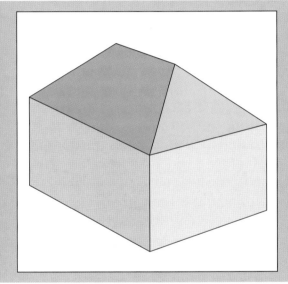

Hip A hip roof (like the one *shown here*) slopes upward from all walls of the structure to the top. It's often used when the same overhang is desired all the way around a structure. Although hip roofs are challenging to frame, they eliminate the gable ends, which reduces the amount of exterior wall that requires maintenance. One common variation of the hip roof is the Dutch hip or modified hip roof; this style provides for louvers on the ends, making it easier to vent.

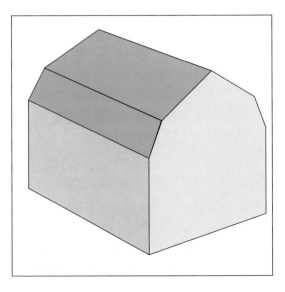

Gambrel A gambrel roof is a type of gable roof where two different slopes combine to create a unique look (*see the drawing at left*). The lower slope of a gambrel roof is generally much steeper than the upper slope; this creates considerably more living space on the second floor. It's also common to see a gambrel roof that features a series of dormers. Since this design is so practical, gambrel roofs are often used for barns and other storage buildings.

Blueprints and Plans

Originally, a blueprint was a copy of a plan that had a deep blue background with white lines. Nowadays, they're referred to as prints or plans and are typically black and white. Plans are drawn to scale and are the main form of communication between the architect, builder, homeowner, and construction professionals. It's important for the homeowner or do-it-yourselfer to be able to read and understand these plans.

Although not everyone can look at a set of plans and "see" the structure, addition, or proposed work, knowing the abbreviations, symbols and types of plan views will certainly help in this endeavor (*see the chart below*). Architects often find it convenient to use abbreviations on blueprint drawings to save space. Some of the most common abbreviations are listed on *page 21;* note that only capital letters are used.

Symbols are also used to identify common items such as a stove, sink, or window. These may be intuitive (like the symbol for a bathtub) or not—like most of the plumbing and electrical symbols. *See page 23* for a drawing illustrating some of the most commonly used symbols.

A full set of house plans will call for a site plan (or plot plan), a foundation plan, a floor plan (*see page 22*), a structural (or framing) plan, and a set of elevations. The plot plan gives the overall view from above and shows the basic shape of the structure and how it sits on the property; it's used mostly by a general contractor to make sure the house ends up in the correct spot. The foundation plan is used by a concrete contractor to locate and install footings, piers, and walls. The floor plan shows the horizontal surface and provides information on the size and arrangement of all the rooms, including electrical, plumbing, and heating fixture placement. Some plans include a separate framing plan to make the floor plan easier to read. Elevations show a view of the structure from the side to help the homeowner visualize the structure.

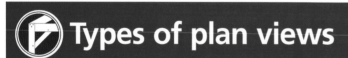

Types of plan views

Type of View	Description
Plan or Overhead	A two-dimensional representation of what the structure will look like when viewed from overhead
Elevation	A two-dimensional view of what the structure will look like from the front, back, or either side
Section	An elevation of what the structure would look like if you were to slice through it at a specified point
Projection	A three-dimensional view of the structure viewed at an angle; you'll typically see at least three sides of the structure at once

Common blueprint abbreviations

Term	Abbrev.	Term	Abbrev.	Term	Abbrev.
Aluminum	ALUM	Floor	FL	Retaining wall	RW
Anchor bolt	AB	Footing	FTG	Ridge	RDG
Basement	BSMT	Foundation	FDN	Riser	R
Bathroom	BATH	Furnace	FURN	Roof	RF
Bathtub	BT	Gauge	GA	Roofing	RFG
Beam	BM	Girder	GDR	Room	RM
Bedroom	BR	Glass	GL	Rough opening	RO
Block	BLK	Grade	GR	Screen	SC
Board	BD	Ground	GRND	Sewer	SEW
Brick	BRK	Gypsum board	GYP BD	Shake	SHK
Building	BLDG	Hardboard	HBD	Sheathing	SHTH
Building line	BL	Hardwood	HWD	Shingel	SHGL
Cabinet	CAB	Heat	H	Shower	SH
Casement	CSMT	Hose bib	HB	Siding	SDG
Cedar	CDR	Insulation	INSUL	Sill	SL
Ceiling	CLG	Interior	INT	Sink	SK
Center	CTR	Jamb	JMB	Skylight	SKL
Centerline	CL	Joist	JST	Sliding door	SL DR
Chimney	CHIM	Kitchen	KIT	Soffit	SOF
Closet	CLOS	Laundry	LAU	Soil pipe	SP
Column	COL	Lavatory	LAV	Solar panel	SLR PAN
Concrete	CONC	Light	LT	South	S
Cornice	COR	Linen closet	LC	Stack vent	SV
Detail	DET	Living room	LR	Stairs	ST
Diameter	DIAM	Louver	LV	Stairway	STWY
Dining room	DR	Medicine cabinet	MC	Steel	STL
Dishwasher	DW	Metal	MET	Top-hinged	TH
Door	DR	North	N	Tread	TR
Dormer	DRM	On center	OC	Utility room	UR
Double-hung	DH	Opening	OPNG	Ventilation	VENT
Douglas fir	DF	Overhang	OH	Vent stack	VS
Downspout	DS	Panel	PNL	Vinyl tile	V TILE
Drain	DR	Partion	PTN	Water	W
Drywall	DW	Plate	PL	Water closet	WC
East	E	Plywood	PLYWOOD	Waterproof	WP
Electric	ELEC	Porch	P	West	W
Elevation	EL	Pressure-treated	PT or P/T	Wide flange	WF
Exterior	EXT	Rafter	RFTR	White pine	WP
Finish	FIN	Redwood	RWD	Window	WDW
Fireplace	FPL	Refrigerator	REF	Wood	WD
Fixture	FIX	Reinforced	REIN	Yellow pine	YP

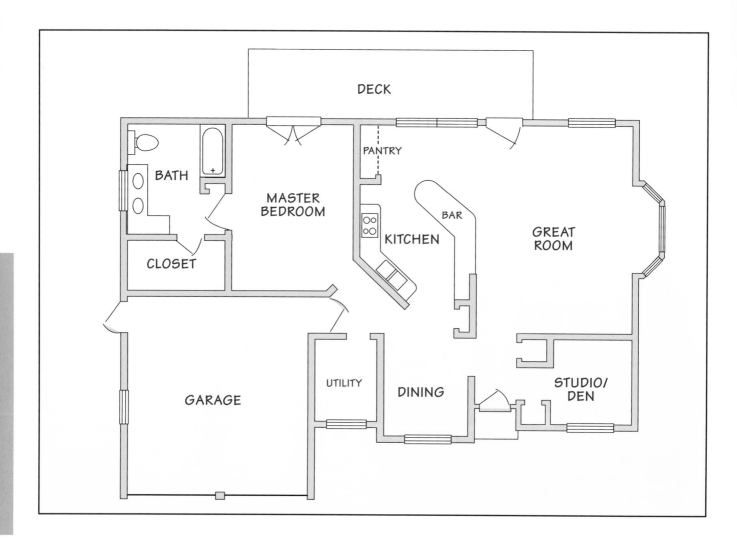

The image shows a floor plan with the following room labels: DECK, BATH, MASTER BEDROOM, PANTRY, BAR, GREAT ROOM, KITCHEN, CLOSET, GARAGE, UTILITY, DINING, STUDIO/DEN.

Floor Plan Example

When most homeowners think of a blueprint or plan, they visualize a floor plan like the one *shown above*. The reason for this is that a floor plan illustrates the size and arrangement of the rooms in a home in a single glance. Add to this the fact that you can shoehorn a surprisingly large amount of information into one of these drawings, and it's no wonder it's so popular. If the structure is to have multiple stories, a separate floor plan will be drawn up for each level.

To help homeowners make intelligent design decisions, many builders and architects have computer programs that can shown a 3-D representation of a structure, an addition, or even a remodeling job. Some are so sophisticated that you can "walk" through the structure to get an even better idea of how the finished project will look.

SYMBOLS

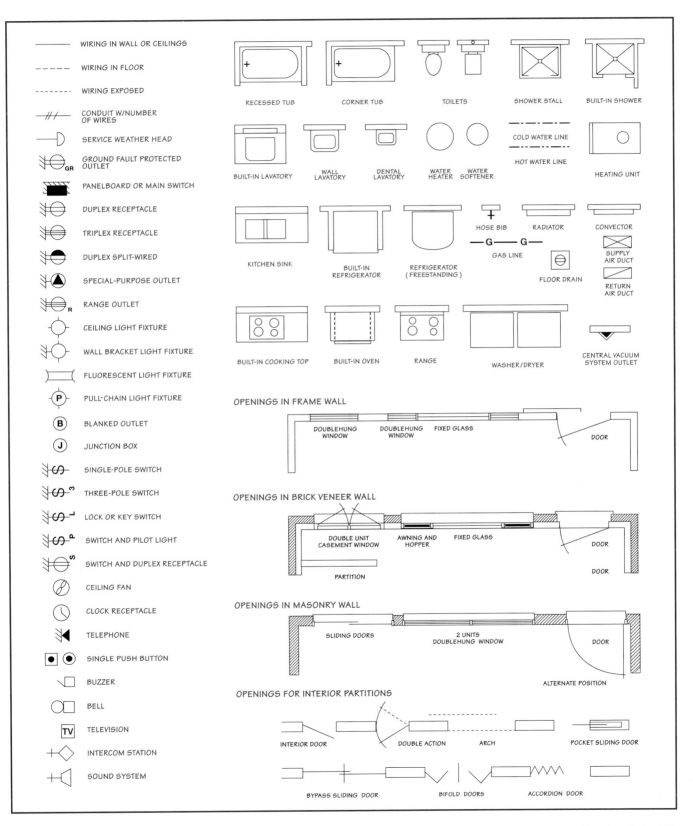

WIRING IN WALL OR CEILINGS

WIRING IN FLOOR

WIRING EXPOSED

CONDUIT W/NUMBER OF WIRES

SERVICE WEATHER HEAD

GROUND FAULT PROTECTED OUTLET

PANELBOARD OR MAIN SWITCH

DUPLEX RECEPTACLE

TRIPLEX RECEPTACLE

DUPLEX SPLIT-WIRED

SPECIAL-PURPOSE OUTLET

RANGE OUTLET

CEILING LIGHT FIXTURE

WALL BRACKET LIGHT FIXTURE

FLUORESCENT LIGHT FIXTURE

PULL-CHAIN LIGHT FIXTURE

BLANKED OUTLET

JUNCTION BOX

SINGLE-POLE SWITCH

THREE-POLE SWITCH

LOCK OR KEY SWITCH

SWITCH AND PILOT LIGHT

SWITCH AND DUPLEX RECEPTACLE

CEILING FAN

CLOCK RECEPTACLE

TELEPHONE

SINGLE PUSH BUTTON

BUZZER

BELL

TELEVISION

INTERCOM STATION

SOUND SYSTEM

RECESSED TUB

CORNER TUB

TOILETS

SHOWER STALL

BUILT-IN SHOWER

BUILT-IN LAVATORY

WALL LAVATORY

DENTAL LAVATORY

WATER HEATER

WATER SOFTENER

COLD WATER LINE

HOT WATER LINE

HEATING UNIT

KITCHEN SINK

BUILT-IN REFRIGERATOR

REFRIGERATOR (FREESTANDING)

HOSE BIB

RADIATOR

CONVECTOR

GAS LINE

FLOOR DRAIN

SUPPLY AIR DUCT

RETURN AIR DUCT

BUILT-IN COOKING TOP

BUILT-IN OVEN

RANGE

WASHER/DRYER

CENTRAL VACUUM SYSTEM OUTLET

OPENINGS IN FRAME WALL

DOUBLEHUNG WINDOW

DOUBLEHUNG WINDOW

FIXED GLASS

DOOR

OPENINGS IN BRICK VENEER WALL

DOUBLE UNIT CASEMENT WINDOW

AWNING AND HOPPER

FIXED GLASS

DOOR

DOOR

PARTITION

OPENINGS IN MASONRY WALL

SLIDING DOORS

2 UNITS DOUBLEHUNG WINDOW

DOOR

ALTERNATE POSITION

OPENINGS FOR INTERIOR PARTITIONS

INTERIOR DOOR

DOUBLE ACTION

ARCH

POCKET SLIDING DOOR

BYPASS SLIDING DOOR

BIFOLD DOORS

ACCORDION DOOR

Permits and Codes

Although most folks think that building permits, codes, and the good folks who enforce them (local building inspectors) are a nuisance, they're all there to protect you and future buyers of your home. Without codes, a building could be made with inferior materials, poor construction methods, and unsafe fixtures. Codes protect you from unscrupulous builders and even from yourself. Even with the best intentions, many do-it-yourselfers can alter their homes to make them unsafe by tackling a home improvement job without consulting the local building inspector. Yes, it does take some added effort, and there's usually a small expense involved, but your family's health and safety are worth it.

There are a number of framing improvements that absolutely must be cleared with your local inspector: removing, altering, or installing a door or window in a load-bearing wall; removing or altering a load-bearing wall; and any job where you'll be altering ceiling or floor joists, rafters, or the roof. Even if you're positive that a wall you want to remove is a non-load-bearing wall, call in an inspector to take a look and confirm that it is or isn't what you think it is. Seriously, if you remove a load-bearing wall in a structure, it can and will cave in over time.

Although there are numerous building codes used throughout the United States—the Uniform Building Code, the Standard Building Code, and the National Building Code to name a few—there is no universally accepted code. Most municipalities adopt one of these codes and then modify it to meet their needs. What's code in one town may not be code in the neighboring town. When in doubt (and even if there is no doubt), double-check with your local building department to make sure your home improvement work is being done to code.

BOROUGH OF EMMAUS

APPLICATION FOR PERMIT

AS REQUIRED BY BOROUGH ZONING ORDINANCE No. 539

Application is hereby made for a permit to erect or alter a structure which shall be located as shown on diagram attached to this sheet and/or to use the premises for the purposes described herewith. It is understood and agreed by this applicant that any error, misstatement or misrepresentation of material fact, either with or without intention on the part of this applicant, such as might or would operate to cause a refusal of this application, or any change in the location, size or use of structure or land made subsequent to the issuance of a permit, without approval of the Zoning Officer, shall constitute sufficient ground for the revocation of permit.

A. LOCATION, OWNERSHIP AND PRESENT USE OF PROPERTY:
1. Street and Number _____
2. Deed Owner _____
3. Owner's Address _____
4. Present Tenant _____ Owner's Consent for proposed work? _____
5. Present Use of Structure, Building, or Land _____
6. Site is located in _____ District as shown on ZONING MAP dated 4 | 17 | 78

B. PROPOSED USE OF STRUCTURE AND/OR LAND: _____ AMENDED _____
NEW STRUCTURE ☐   ADDITION ☐   LAND USE ☐   SIGN ☐   OCCUPANCY ☐
ACCESSORY USE ☐   CHANGE OF USE ☐   FENCE ☐   DRIVEWAY ☐   OTHER _____
1. Proposed Use of Structure, Building, or Land _____
2. Description of Work _____

3. Plan is attached hereto. Yes ☐ No ☐   ESTIMATED CONSTRUCTION VALUE $ _____

C. REQUEST FOR OCCUPANCY:
1. Desired Use _____ Former Use, if any _____

D. APPLICANT:
1. Name of Applicant _____
2. Address of Applicant _____
3. Owner, Lessee, or authorized agent for owner of subject property _____
4. Applicant's Signature _____

E. APPROVAL AND DATES OF ACTION TAKEN:
1. Application approved. Yes ☐ No ☐   Date _____
2. Reason for Denial of Application _____

_____ Zoning Officer
3. Applied to Zoning Hearing Board _____ 19 ___ APPEAL: Yes ☐ No ☐ Hearing No. _____
4. Special Exception ☐   Variance ☐
5. Board's Decision. Granted ☐   Denied ☐   Date _____
ORDER: _____

F. ISSUANCE OF PERMIT:

FEE $ _____ TYPE _____ DATE ISSUED _____ NO. _____
OCCUPANCY PERMIT WILL BE REQUIRED UPON COMPLETION OF WORK. ORD. #444 ☐

Recommended rafter spans for roof slope 3 in 12 or less: 40# snow load, 10# dead load

Species	Grade	2×8, 16" OC	2×8, 24" OC	2×10, 16" OC	2×10, 24" OC	2×12, 16" OC	2×12, 24" OC
Douglas fir - Larch	1	14' 5"	11' 9"	17' 8"	14' 5"	20' 5"	16' 8"
	2	13' 6"	11' 0"	16' 6"	13' 6"	19' 2"	15' 7"
	3	10' 3"	8' 4"	12' 6"	10' 2"	14' 6"	11' 10"
Douglas fir - South	1	13' 8"	11' 2"	16' 9"	13' 8"	19' 5"	15' 10"
	2	13' 1"	10' 9"	16' 0"	13' 1"	18' 7"	15' 2
	3	9' 11"	8' 2"	12' 2"	9' 11"	14' 1"	11' 6"
Hem-Fir	1	14' 1"	11' 6"	17' 2"	14' 0"	19' 11"	16' 3"
	2	13' 4"	10' 10"	16' 3"	13' 3"	18' 10"	15' 5"
	3	10' 3"	8' 4"	12' 6"	10' 2"	14' 6"	11' 10"
Spruce-Pine Fir (South)	1	13' 4"	10' 10"	16' 3"	13' 3"	18' 10"	15' 5"
	2	12' 6"	10' 3"	15' 3"	12' 6"	17' 9"	14' 6"
	3	9' 5"	7' 8"	11' 6"	9' 5"	13' 4"	10' 11"
Western Woods	1	11' 8"	9' 6"	14' 3"	11' 7"	16' 6"	13' 6"
	2	11' 8"	9' 6"	14' 3"	11' 7"	16' 6"	13' 6"
	3	8' 10"	7' 3"	10' 10"	8' 10"	12' 6"	10' 3"

Data for both charts courtesy of Western Wood Products Association

Recommended floor joist spans: 40# live load, 10# dead load

Species	Grade	2×8, 16" OC	2×8, 24" OC	2×10, 16" OC	2×10, 24" OC	2×12, 16" OC	2×12, 24" OC
Douglas fir - Larch	1	13' 1"	11' 0"	16' 5"	13' 5"	19' 1"	15' 7"
	2	12' 7"	10' 3"	15' 5"	12' 7"	17' 10"	14' 7"
	3	9' 6""	7' 9"	11' 8"	9' 6"	13' 6"	11' 0"
Douglas fir - South	1	12' 0"	10' 5"	15' 3"	12' 9"	18' 1"	14' 9"
	2	11' 8"	10' 0"	14' 11"	12' 2"	17' 4"	14' 2"
	3	9' 3"	7' 7"	11' 4"	9' 3"	13' 2"	10' 9"
Hem-Fir	1	12' 7"	10' 9"	16' 0"	13' 1"	18' 7"	15' 2'
	2	12' 0"	10' 2"	15' 2"	12' 5"	17' 7"	14' 4"
	3	9' 6"	7' 9"	11' 8"	9' 6"	13' 6"	11' 0"
Spruce-Pine Fir (South)	1	11' 8"	10' 2"	14' 11"	12' 5"	17' 7"	14' 4"
	2	11' 4"	9' 6"	14' 3"	11' 8"	16' 6"	13' 6"
	3	8' 9"	7' 2"	10' 9"	8' 9"	12' 5"	10' 2"
Western Woods	1	10' 10"	8' 10"	13' 3"	10' 10"	15' 5"	12' 7"
	2	10' 10"	8' 10"	13' 3"	10' 10"	15' 5"	12' 7"
	3	8' 3"	6' 9"	10' 1"	8' 3"	11' 8"	9' 6"

Chapter 2
Tools and Materials

Although you could frame a wall with a few materials (some studs and nails) and a couple of hand tools (a hammer and a saw), there's a whole lot more out there to choose from. New materials such as metal framing and plywood I-beams are lighter and stronger than their solid-wood cousins. And there are tools and accessories available that can add a new level of accuracy to your work.

In this chapter, I'll start with the tools that you'll need for most framing jobs, beginning with general-purpose tools such as those needed for demolition work, measuring, and layout (*opposite page*); cutting tools; power tools; and safety gear (*page 28*)—everything from plumb bobs, for accurately locating framing members, to dust masks to protect your lungs from sawdust and demolition dust. Then I'll go over some specialty tools that you might find worth the investment (*page 29*): specialty tools for adding accuracy to your cuts, such as clamp-on straightedges and rip fences. I also discuss work surface and storage accessories that effectively let you bring your shop to the work site—folding workbenches, "bucket" organizers, and tool belts, to name just a few.

Next, I'll describe in detail the materials that you have to choose from for your framing project: dimension lumber (*page 30*); metal framing (*page 31*); and sheet stock such as plywood, particleboard, and OSB (oriented-strand board) (*pages 32–33*). This is followed by a section on engineered boards like plywood I-joists, box beams, and laminated-veneer lumber (LVL) (*pages 34–35*). There's also information on common materials used to seal and insulate a wall from the elements: moisture barriers and insulation (*page 35*).

Fasteners are the topic of the rest of the chapter. It's extremely important to understand the differences among the huge variety of fasteners available so that you can match the correct fastener to the job. This is critical in order to make sure walls, ceilings, and floors stay put and will safely support their designated loads over time. I'll take you through the array of fasteners, starting with nails (*pages 36–37*), screws (*page 38*), metal framing connectors such as anchors and post caps (*page 39*), and joist hangers (*page 40*). Finally, I'll cover the most common adhesives and caulks used for framing jobs (*page 41*).

General-Purpose Tools

Demolition Many of the framing jobs you'll tackle will require some demolition work—tearing out old flooring, removing cabinets or a small section of a wall. You'll find the following tools useful for this type of work (*from left to right*): screwdrivers for general dismantling; a sledge-hammer for persuading stubborn walls to come down; a pry bar for pulling out boards and fixtures; a cold chisel or set of inexpensive chisels for chopping out holes in walls or flooring; a claw hammer for general removal; and a cat's paw for removing nails flush or below the surface of a workpiece.

Measuring One of the most critical steps in any framing job is measuring and laying out the grid or starting point. The tools *shown* should be in every homeowner's toolbox: a 25-foot tape measure; a framing square to check for perfect right angles; a folding rule for short accurate measurements; a combination square to check for right angles; and a speed square to take quick, short measurements.

Layout In addition to measuring, laying out intended work is critical to the success of any framing job you take on. The following tools (*shown*) should also be in your toolbox: a 4-foot-long level and a shorter torpedo level for checking framing members for level and plumb; a compass to draw circles and arcs; a contour gauge for laying out odd shapes; a plumb bob and string for transferring location points for vertical framing members (such as from ceiling to floor); and a chalk line for striking long layout lines.

Cutting Tools Framing requires only a couple of specialized tools for cutting. Although you can cut dimension lumber and trim it with a hand saw, it'll be a long day (or days, most likely). If you don't own a power miter saw or "chop" saw , consider renting one or borrowing one from a friend. In addition to this, you'll need a wood chisel and a block plane for fine-tuning the fit of parts; a compass saw for cutting notches in framing members, flooring, and cabinets; a pocket or utility knife to trim shims, etc.; and a drywall saw for "cutting in" electrical boxes.

Power Tools Power tools can make quick work of many of the tedious tasks associated with installing framing. *Shown clockwise from top right:* a cordless drill with a ⅜" chuck for smaller-diameter holes; a saber saw for cutting access holes; an electric drill with a ½" chuck for large-diameter holes; a reciprocating saw for demolition work; a cordless trim saw for straight-square cuts; and a right-angle drill for tight spots.

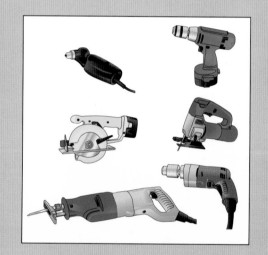

Safety Gear As with any home improvement work, it's important to protect yourself by wearing appropriate protective gear. Keep the following on hand (*clockwise from bottom left*): leather gloves to protect your hands; safety goggles to protect your eyes; knee pads not only to cushion your knees but also to protect them; ear muffs or plugs for when working with power tools; and a dust mask or respirator to protect your lungs from sawdust and the dust raised during demolition.

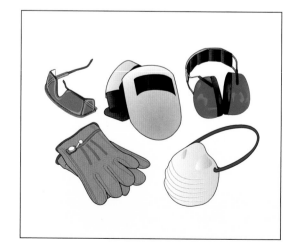

Specialty Tools

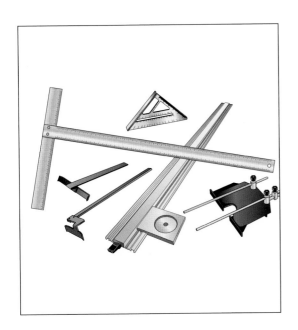

Accurate Cutting For a wall or structure to go up level and plumb, care must be taken to ensure that the framing members and other materials that you use are cut square. There are a number of accessories that can help with this. *Shown clockwise from top left:* a T-square for laying out and cutting drywall; a speed square that acts as a guide for a circular saw on short cuts; an edge guide for a router for cutting rabbets, dadoes, and grooves; a clamp-on metal straightedge that's used to guide either a router or a circular saw for accurately cutting sheet goods; and a pair of slide-on fences for ripping boards to width with either a circular saw or a saber saw.

Work Surfaces and Organizers Since you have to take your tools to the job when it comes to a framing task, you'll want to minimize the number of tool-toting trips back and forth from the shop to the job site. Here's where portable work surfaces and tool organizers come in handy. *Shown clockwise from top left:* the ever popular sawhorse to temporarily support a workpiece for a cut; a "bucket" organizer that slips into an ordinary 5-gallon bucket and holds a surprisingly large amount of tools; a sturdy tool belt with hammer loops to keep fasteners and your favorite tools right at hand; a portable folding workbench; and a tool bag for longer tools (such as a reciprocating saw or a worn-drive saw).

Dimension Lumber

Because most softwood lumber is used in construction, it is cut to standard sizes. This way architects and builders around the world can all use the same "building blocks" when they design or build a structure. The length of a softwood board is given in actual dimension, and the width and thickness are given in "nominal" dimensions; actual dimensions are somewhat less. Nominal dimensions are based on rough-cut, green lumber; actual dimensions describe boards after they've been dried and surfaced on all four sides.

Although softwood lumber can be cut in 1-foot increments, 2-foot increments are more common. The width of softwood lumber varies from 2" to 16", usually in 2" increments. Thickness is generally categorized into three groups: *boards* are

lumber that's less than 2" in thickness, *dimension lumber* ranges from 2" to 5", and *timbers* are more than 5" (*see the chart below left*).

When dimension lumber is graded or sorted by its characteristics, much more than appearance comes into play: Strength, stiffness, and other mechanical properties are all taken into consideration. The problem is, no two woods have identical characteristics. This means that every softwood species has its own set of grading guidelines.

There are four grades of structural light framing lumber: Select structural, No. 1, No. 2, and No. 3.

Select structural is the highest grade in structural light framing and is recommended where appearance is as important as strength and stiffness.

When appearance is still important but is secondary to strength, No. 1 grade is the best choice. No. 2 structural light framing lumber is recommended for general construction. When strength is not a factor, No. 3 grade can be used.

Standard sizes of dimension lumber

Item	Nominal Thickness (inches)	Dressed Thickness (inches)	Dressed Thickness (mm)	Nominal Width (inches)	Dressed Width (inches)	Dressed Width (mm)
Boards	1	¾	19	2	1½	38
Boards	1¼	1	25	4	3½	89
	1½	1¼	32	6	5½	140
				8	7½	184
				10	9¼	235
				12	11¼	286
Dimension	2	1½	38	2	1½	38
	2½	2	51	4	3½	89
	3	2½	64	6	5½	140
	3½	3	76	8	7½	184
	4	3½	89	10	9¼	235
	4½	4	102	12	11¼	286

Metal Framing

Non-Load-Bearing Studs Steel studs are becoming increasingly popular as a replacement for wood studs. They're easy to use, are perfectly straight, and cost about the same as wood studs. Steel studs are available in two common grades or categories: load-bearing (LB) and non-load-bearing (NLB). NLB drywall studs are typically made of 25-gauge steel and should be used only for partition walls (*see the photo at left*). LB studs use a heavier gauge to handle the additional stress (*see below*).

Load-Bearing Studs Load-bearing studs can usually be identified by the bent lip at the edge of the flange (*see the photo at left*). This added lip helps keep the stud rigid so that it can support heavier loads. Most manufacturers of LB studs also make track without lips so that the studs can slip in place without any cutting being necessary. These tracks are the equivalent of top and sole plates. Studs and tracks are available in widths ranging from 1⅝" to 6". (*See pages 58–59 for information on working with metal framing.*)

PRECUT ACCESS HOLES

Some brands of metal framing have pre-punched knockouts for running electrical lines. Because the edges of the knockouts are razor-sharp, the framing manufacturers sell plastic bushings that easily snap into place to protect the wiring. Insert a bushing into each side of the metal stud, and press the pieces together until they snap together. Note: If you're planning on using these cutouts to run wiring, take care as you cut the studs to length (and position them) so that the cutouts line up horizontally. This will simplify pulling the cable through the studs.

Sheet Stock

Plywood

Most softwood plywood is manufactured for use in either industrial or construction applications. That's why most standards for softwood plywood deal exclusively with how it must perform in a designated application rather than from what or how the plywood is manufactured. Certain grades of softwood plywood, however, are quite suitable for woodworking projects where appearance isn't critical or if the plywood will be used as a base for veneer, laminate, or paint.

Grade in softwood plywood generally refers to the quality of the veneer used for the face and back veneers (A-B, B-C, etc.); *see the chart below.* Grade can also refer to the intended end use of the panel, such as Sheathing, or Underlayment. The standard that most softwood plywood manufacturers adhere to is Voluntary Product Standard PS 2-92, Performance Standard for Wood-Based Structural-Use Panels, published by the APA, now called the Engineered Wood Association.

Exposure Durability

For projects that will be subjected to moisture such as outdoor play equipment, sheds, or boat-building, there are three exposure durability classifications you should be familiar with: Exterior, Exposure 1, and Exposure 2. Exterior panels have a fully waterproofed bond and are designed for applications subject to permanent exposure to moisture. Exposure 1 panels should be used for protected applications where the glue bond must be waterproof. Exposure 2 panels are intended for protected construction and industrial applications. Panels rated as Interior are manufactured with interior glue and should be used only in interior applications.

Softwood plywood grades

Veneer Grade	Characteristics
A	Smooth, paintable. Not more than 18 neatly made repairs permitted that are boat, sled, or router type, and parallel to grain. Wood or synthetic repairs permitted. May be used for natural finish in less-demanding applications.
B	Solid surface. Shims, sled, or router repairs, and tight knots to 1" across grain, permitted. Wood or synthetic repairs permitted. Some minor splits permitted.
C (plugged)	Improved C veneer with splits limited to ⅛" width and knotholes or other open defects limited to ¼" × ½". Admits some broken grain. Wood or synthetic repairs permitted.
C	Tight knots to 1½". Knotholes to 1" across grain and some to 1½" if total width of knots and knotholes is within limits. Synthetic or wood repairs, and discoloration and sanding defects that do not impair strength permitted. Limited splits allowed. Stitching permitted.
D	Knots and knotholes to 2½" width across grain and ½" larger within limits. Limited splits allowed. Stitching permitted. Limited to Interior, Exposure 1, and Exposure 2 panels.

Particleboard Particleboard is a wood-panel product that's produced mechanically: Wood is reduced into small particles, adhesive is applied to the particles, and then heat and pressure turn a mat of particles into a panel product. Particleboard is heavy (a ¾"-thick sheet weighs almost 100 pounds), flat, fairly stable, and inexpensive. Although there are a number of grades available, you'll probably find just two of these at your local lumberyard or building center: underlayment and industrial. Underlayment particleboard is commonly used for floor sheathing. Industrial-grade has a core made of coarse particles sandwiched between two outer layers of finer particles, which creates a smoother, flatter surface that's ideal as a substrate for countertops.

OSB Oriented-strand board (OSB) is an engineered panel made from strands of wood bonded together with a waterproof resin under heat and pressure. It has found wide acceptance in the construction industry, where it's used primarily for roof, wall, and floor sheathing. OSB is made in three grades: APA-Rated Sheathing, APA-Rated Sturd-I-Floor, and APA-Rated Siding (*see the chart below*). Along with a grade, you'll always find a span rating as part of the grade stamp. A span rating is the recommended center-to-center spacing of supports, in inches, over which the panels should be installed. A span rating looks like a fraction (such as 32/16), but it isn't. The left-side number describes the maximum spacing of supports in inches when the panel is used for roof sheathing; the right-side number denotes the maximum spacing of supports when the panel is used for subflooring.

Grades of OSB

Grade	Thicknesses	Application
APA-rated sheathing	⁵⁄₁₆", ³⁄₈", ⁷⁄₁₆", ¹⁵⁄₃₂", ¹⁄₂", ¹⁹⁄₃₂", ⁵⁄₈", ²³⁄₃₂", and ³⁄₄"	Can be used for subflooring; wall sheathing; and industrial applications such as shelving, furniture, trailer liners, and recreational vehicle floors, roofs, and components
APA-rated flooring (Sturd-I-Floor)	¹⁹⁄₃₂", ⁵⁄₈", ²³⁄₃₂", ³⁄₄", ⁷⁄₈", 1", and 1⅛"	Intended for use as single-layer flooring under carpet and pad (most flooring has tongue-and-groove edges).

Engineered Beams

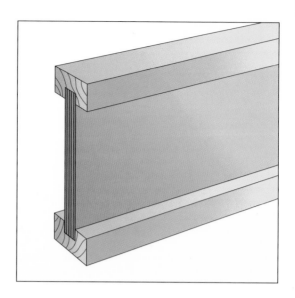

Plywood Joists Plywood joists or wood I-beams are becoming increasingly popular in construction, and for good reason: They're light, easy to use, and perfectly straight, and they can handle long spans (*see the drawing at right*). They're made by sandwiching a vertical piece of plywood or OSB called the web between two strips of grooved lumber referred to as rails. They come in four heights: $9\frac{1}{2}$", $11\frac{7}{8}$", 14", and 16". You can nail wood I-beams in place just as you would a standard beam.

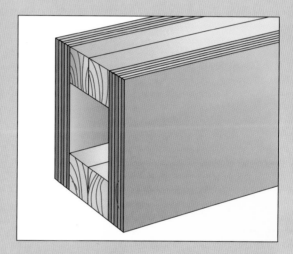

Box Beams Box beams, also called glued-plywood-and-lumber beams, are made by sandwiching 2-by lumber between two plywood outer panels called skins (*see the drawing at right*). They're used for supporting structural loads and must be designed by a qualified architect or engineer. The materials used must be rated for their intended use to fully support the load. For maximum strength, the plywood should be both glued and nailed at all edges, and plywood joints or seams must be offset.

LAMINATED BEAMS

There are a number of laminated beam products available that are useful both as support beams and as headers. Laminated-veneer lumber (LVL) is made by laminating vertical strips of veneer together; it comes in widths in increments of $1\frac{3}{4}$". Two pieces nailed together match the thickness of a 2×4 wall. Heights of LVL beams range from $5\frac{1}{2}$" to 18".

Laminated lumber beams, different from LVL, are made by gluing laminated horizontal strips of knot-free lumber together. They're very stable, and each beam is specifically engineered to support the intended building load. They're commonly used for ridge beams, purlins, and floor girders.

Parallel-strand lumber is made by gluing together parallel strands of Douglas fir or Southern yellow pine. These beams are dimensionally stable and won't warp or twist like many of the other engineered beams.

Barriers and Insulation

Vapor and Air Barriers To protect a newly framed exterior wall from moisture, you must cover the wall sheathing with some form of moisture barrier. A moisture barrier is any thin membrane that's applied directly to wall sheathing before siding is installed to prevent moisture from seeping into the wall. Effective barriers stop liquid water but allow water vapor to pass so the walls can "breathe." Air barriers limit air filtration through a wall and may be a part of the insulation applied to the wall.

Rigid Insulation Rigid insulation, or foam board, is often used to increase the thermal performance of a wall. It can be applied directly over sheathing before the siding is installed. Or it may be installed to the inside of a wall *as shown in the photo at left.* Rigid foam is commonly available in thicknesses ranging from 1/2" to 2" and comes in 4×8-foot sheets or else trimmed into panels designed to fit common stud-spacing patterns. When applied to an interior wall, it is often glued in place with construction adhesive.

Fiberglass Fiberglass insulation has long been the favorite in construction. It cuts easily, goes up quickly, and has admirable insulation properties. Fiberglass insulation comes in faced or unfaced rolls or precut lengths. The facing used may be paper or a vapor barrier—both versions have tabs that are folded over the studs and stapled in place. When unfaced batts are installed in walls, a continuous vapor barrier should be used to cover the studs and insulation—this system actually provides a better barrier than faced insulation, since the barrier is continuous.

Fasteners: Nails

The primary fastener used by the framer is the nail, and there is a wide variety of nail types to choose from. Each type of nail listed in the chart *below* or shown in the drawing on the bottom of the *opposite page* is designed for a specific purpose. Common nails are by far the most popular for framing: They're the beefiest, they have the least tendency to bend, and their broad head is easy to strike with a hammer. Other nails offer a variety of shank type, head design, and holding power.

Nails are sized using the antiquated "penny" system (abbreviated as "d"). This system is based on how much 100 nails used to cost (talk about inflation!). Penny is now used to indicate the length of the nail; *see the chart on the opposite page.* You can purchase nails in 1-, 5-, and 50-

pound boxes, or in any quantity from a retailer that stores their nails in open bins.

If the nails you'll be using will be exposed to the elements (such as when building a deck, porch, or shed, or installing a roof), they should either be hot-dipped galvanized or electroplated galvanized, or be made of a metal that's impervious to moisture, such as aluminum or stainless steel.

Typical nails used in construction

Type	Application	Design
Box Nails	General construction; light-duty use	Thin shaft is less likely to split wood, but it bends more easily
Brads	Thin trim and plywood edging	Sized by wire gauge, from 20 ga. to 15 ga.; 20 is the thinnest
Casing Nails	Interior and exterior trim	Similar to finish nails but with a larger head for better holding power
Cement-Coated Sinkers	General construction	Coated with cement adhesive for better holding power
Common Nails	General framing; heavy-duty use	Thick shanks split wood, but don't bend easily
Finish Nails	Interior and exterior trim	Smaller head can be set below the finish surface

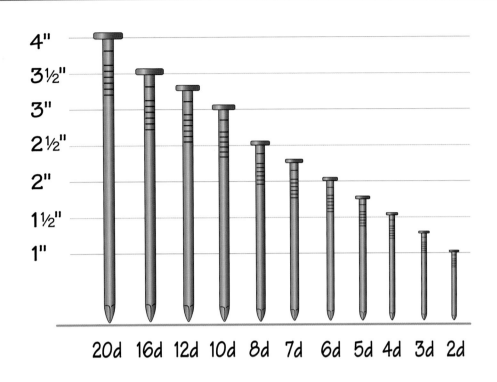

4"
3½"
3"
2½"
2"
1½"
1"

20d 16d 12d 10d 8d 7d 6d 5d 4d 3d 2d

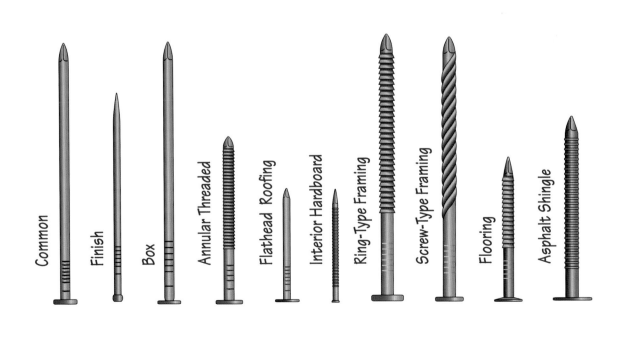

Common   Finish   Box   Annular Threaded   Flathead Roofing   Interior Hardboard   Ring-Type Framing   Screw-Type Framing   Flooring   Asphalt Shingle

Fasteners: Screws

Wood Screw Tapered wood screws, long the fastener of choice for woodworkers, require a pilot hole that's either tapered or of two different diameters. One type of specialty drill bit, a pilot bit set, can drill both diameter holes at the same time along with a countersink or counterbore to leave the screw head flush or below the surface of the workpiece. The thicker tapered shanks of wood screws are stronger than straight-shank screws and are often used to secure the hinges of heavy doors to doorjambs.

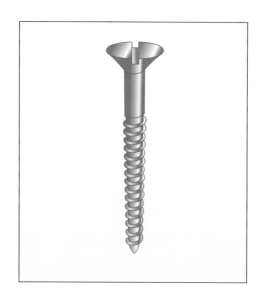

Straight-Shank Straight-shank screws are becoming increasingly popular because they require only a single-diameter pilot hole. This type of screw, often referred to as a drywall screw, comes with either fine or coarse threads. Coarser threads drive faster into wood and afford a solid grip. Straight-shank screws can be coated or uncoated and, when galvanized, are useful for outdoor projects, such as decks, porches, and sheds.

Square-Drive A version of the straight-shank screw, a square-drive screw features a square recess in the head, which offers a more positive drive system. Square-drive screws require a special square-tipped screwdriver or screwdriver bit. Some screws have a combination recess that will accept either a square-drive tip or a Phillips-head screwdriver. Square-drive screws are particularly well suited for driving into tough materials or for long screws (such as deck screws), where the enhanced drive system is necessary.

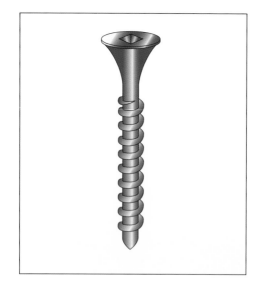

Anchors and Caps

Metal framing connectors are designed for use on 2-by projects where you need to attach framing members together quickly. Connector manufacturers offer an unbelievable assortment of anchors and ties for almost every conceivable application.

Anchors

Many local building codes require framing anchors in numerous situations, such as where floor joists meet headers or rim joists, and where studs meet sole plates; *see the drawing at left for common types.* Make sure to check with your local building inspector to identify what type of anchors (if any) are required in your area. Note: In locations where seismic activity is a possibility, check to see whether code requires special seismic anchors. Seismic anchors are designed to help hold a structure together during an earthquake.

Caps

Another common framing connector is the post or base cap (*see the drawing at left*). The connectors are designed to handle the transition from one framing member to another—quite often the transition from post to beam. A pair of adjustable post caps can be spaced apart to fit a wide variety of framing situations. Retrofit post caps are also installed in pairs and are suitable for heavier loads than adjustable base caps. Retrofit ends caps are similar to post caps except that they're designed for corner support.

Safety Note: All framing connectors are rated to handle a maximum allowable load. If you're not sure about which one to use, check with the manufacturer, a building contractor, or your local building inspector.

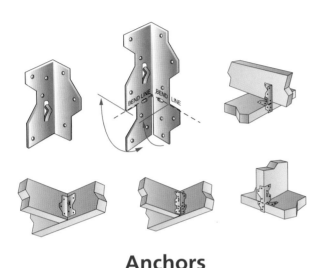

Anchors

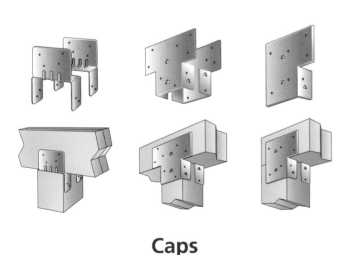

Caps

Illustrations based on original artwork © Simpson Strong-Tie, 2000

Joist Hangers

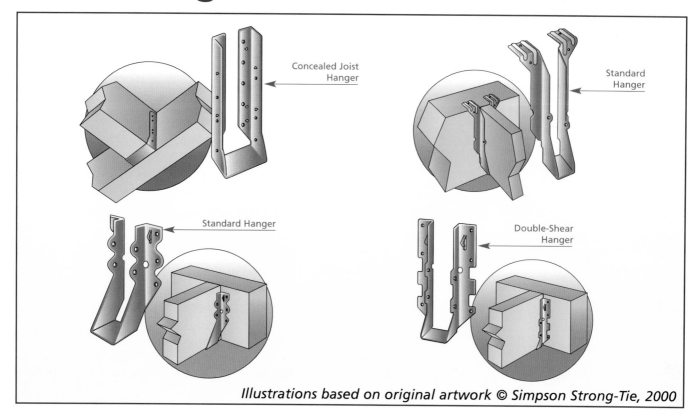

Concealed Joist Hanger

Standard Hanger

Standard Hanger

Double-Shear Hanger

Illustrations based on original artwork © Simpson Strong-Tie, 2000

Joist hangers are the framing connector used by far the most often by both homeowners and professionals. They're designed to support a joist, girder, or other framing member from a post, beam, rim joist, header, etc. Joist hangers create a much stronger joint than is possible by simply nailing framing members together.

Safety Note: In order for joist hangers to be able to support their designated load reliably, they must be installed with special fasteners called joist hanger nails. These nails have a stout shank and increased shear strength.

The drawing *above* illustrates just a few of the most common types of joist hangers available. The standard joist hanger is made of 20-gauge metal and is engineered for strength and economy. Most manufacturers build in prongs that,

when driven into the framing member, will hold the hanger in place long enough for you to drive in nails. They're sized for 2×4s, 2×6s, 2×8s, and 2×10s. Top flange joist hangers have a flange on the top that hooks over the framing member for added support.

Double-shear joist hangers are manufactured from heavier 18-gauge metal and are designed to support heavier loads. Note that with these hangers, the nails must be driven in at an angle to achieve the specified loads. Concealed joist hangers are used for end applications where the hanger flanges need to be concealed for aesthetic reasons. They're made of 14-gauge metal and have extra holes to accept more nails to handle even greater loads.

Caulk and Adhesive

Every framer uses caulk and adhesive. Construction adhesive is what's used most often in framing. It comes in tubes to fit a caulking gun so it can be applied quickly and with some accuracy. It's often used to secure floor sheathing and to secure paneling to furring strips on interior walls.

You'll also find it handy to have two types of caulk on hand for most jobs: acrylic latex and silicone. Acrylic latex is used to fill gaps behind trim prior to painting; silicone is often used in exterior applications such as when caulking around a window or door. *See the chart below for common adhesives and caulks and their applications.*

Caulk and adhesive applications

Type	Category	Design
Acrylic Latex	Caulk	Most common application is filling gaps and voids behind trim; also known as painter's caulk, as it accepts paint well; inexpensive and very easy to use and clean up
Aliphatic Resin	Adhesive	Commonly used to glue up finished trim; also known as carpenter's glue; parts must be held under pressure with clamps to get a good bond
Butyl Rubber	Caulk	Applications where you need to seal metal (such as flashing to masonry); inexpensive, flexible, but very messy to use (wear disposable rubber gloves)
Construction Adhesive	Adhesive	Used to bond floor sheathing to floor joists; panels and rigid insulation to walls; drywall to framing members; etc.; inexpensive; comes in tubes for caulking guns and is available in a waterproof variety
Contact Cement	Adhesive	Most commonly used to bond laminate to countertop surfaces; it is applied to both surfaces, allowed to become tacky, and then joined; positioning is critical, as it bonds on contact
Epoxy Resin	Adhesive	Used to bond together dissimilar materials, such as wood and steel; two-part systems are mixed together and applied; bond is strong but not very flexible
Silicone	Caulk	Wood-to-masonry applications, also bathroom tile joints that won't be painted; extremely flexible and long-lasting but can't be painted
Siliconized Acrylic Latex	Caulk	Interior and exterior applications where a more flexible caulk is needed; bonds better to surfaces and lasts longer than acrylic latex; less messy and easier to use than silicone

Chapter 3
Framing Techniques

I've had a number of opportunities to watch a seasoned carpenter (or a crew) frame a house. It's amazing how efficiently they work. Little effort is wasted, and the structure goes up surprisingly fast. One reason for this is that they really know how to use the basic tools—years of experience have trimmed motions down to the bare essentials. A stud is quickly positioned with a tap of a hammer and a nudge from a foot. Nails are driven in with just a few blows.

Although skills like this develop over a lifetime, they all began with a basic understanding of each tool and how to use it. In this chapter, I'll go over the fundamental techniques that you'll need to take on most interior framing jobs. I'll start with the very important first step of any framing task: layout and measurement. Everything from using the 3-4-5 triangle to guarantee that adjoining walls go up perpendicular to each other (*opposite page*), basic tape measure use (*page 44*), and using squares (*page 45*), to the tools used to check for level and plumb (*page 46*).

Then on to the basic tools and techniques you'll need. Although it may seem obvious, there is a correct way to hold and swing a hammer. The section on *pages 47–48* goes over hammer

fundamentals and includes a number of tips that you'll often see the pros use on site. But there's more to construction than swinging a hammer. Proper nail placement or nailing patterns will ensure that the framing members are firmly attached to each other (and will stay attached over time); *see page 49 for more on this.* In case you've got a lot of framing to do, I've included a section on air nailers (*page 50*) that describes them and their basic use. This is followed by information on *page 51* on how to pull nails for demolition work, or in the course of normal framing when you need to reposition a nail.

Using saws is the topic of the next section: handsaws (*page 52*); circular saws, the workhorse of framing (pages 53–55) including specialty cuts like plunge cuts or cutting sheet stock; the reciprocating saw for demolition work and rough cuts (*page 56*); and the saber saw for curved cuts (*page 57*). Since metal framing is becoming increasingly popular, I've included a section on *pages 58–59* on how to work with it—everything from cutting to assembling. Next, there's a section on how to straighten the inevitable crooked lumber you'll have to work with (*page 60*). And finally, step-by-step directions on how to fine-tune parts for a perfect fit (*page 61*) using a couple of simple hand tools.

Layout and Measurement

An accurate set of reference lines will make any framing job go a lot smoother. To do this, start by measuring and snapping a chalk line where the intended wall or wall section will be. Then use a framing square to lay out a line perpendicular to the first line. Align your chalk string with the line you just drew, and snap a line perpendicular to the first chalk line. To make sure that these two lines are perfectly perpendicular to each other, use the 3-4-5 triangle method described *below*. **Shop Tip:** Since it can be difficult to accurately measure the 5-foot distance between the sides of the triangle with a tape measure, I made a simple jig. It's just a length of wire rope that has two loops at each end. It measures exactly 5 feet from end of loop to end of loop. The advantage of wire rope is that it doesn't stretch like string can. I looped the ends so that I can use an awl to hold one end of the rope in place on the triangle—this makes it easy to check the other end for alignment.

3-4-5 Triangle One of the oldest and most reliable ways to check to make sure reference lines are exactly perpendicular is to use a 3-4-5 triangle. To do this, start by measuring and marking a point 3 feet from the centerpoint where the lines cross (make this mark on either line). Then measure and mark 4 feet from the centerpoint on the adjacent line. Now measure the distance from the 3-foot mark and the 4-foot mark with a tape measure (or a piece of string or wire rope). If the lines are truly perpendicular, the distance will measure exactly 5 feet. If it doesn't, the lines aren't perpendicular, and the position of one of the lines will have to be adjusted.

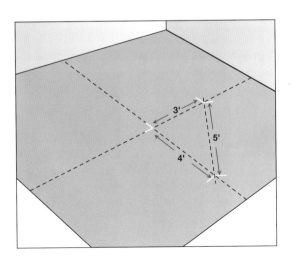

How It Works The 3-4-5 triangle is based on the Pythagorean theorem. This theorem describes the relationship between the sides and angles of a right triangle. In practical terms, it states that if the hypotenuse of a triangle (the longest side) measures 5 feet and the opposite and adjacent sides measure 4 feet and 3 feet, respectively, then the angle that's formed between the adjacent and opposite sides is exactly 90 degrees. If that triangle's hypotenuse doesn't measure exactly 5 feet, then the angle isn't 90 degrees, and some adjustment is necessary.

Using a Tape Measure

Inside Measurements A tape measure will give accurate inside measurements if it has a loose or sliding end hook. This compensates for the hook's thickness so the tape can be used for both inside and outside measurements. For an inside measurement, position the tape parallel to the edge you're measuring, butt it squarely against one surface, and extend the tape until the end hook butts up against the opposite surface (*right*). Read the tape and add the case length to this (it's usually an even measurement, like 3").

Outside Measurements For rough outside measurements, simply engage the end hook of the tape over the edge of the surface to be measured, extend the tape past the opposite surface, and read the tape. If precision is essential, don't use the end hook, as these often get bent out of shape over time, which can result in erroneous measurements. Instead, extend the tape past the surface to be measured so that the 1" mark aligns with the edge *as shown in the photo.* Then extend the tape to the opposite surface and take a reading—don't forget to subtract 1" from this to get the actual length.

AN EXTRA HAND

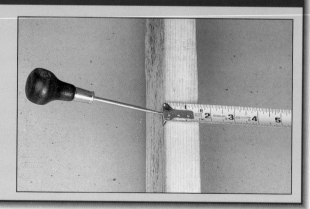

Most tape measures have a hole or holes punched in their end hook. This provides a handy place for inserting an awl to hold one end of the tape in place "hands-free" so that you can take a long measurement by yourself (*see the photo at right*). Most end hooks have holes in both the top and end of the hook so that you can attach the hook from the top or end depending on the surface to be measured. Just make sure that the end of the hook butts firmly up against the edge of the piece to be measured.

Using Squares

Try or Combination Square When you need to mark a square line or check a 90-degree angle, reach for a try or combination square. A try square has a fixed blade; a combination square's blade is adjustable. To use either, you place the head of the tool against an edge of the workpiece. The blade then provides an accurate 90-degree reference. Combination squares are great for laying out parallel lines. Do this by sliding the head along the workpiece with a pencil pressed against the blade, *as shown.*

Framing Square A framing or "carpenter's" square is used for laying out and checking lines for square. To lay out square lines, place the tongue of the square against the edge of the workpiece so it's flush with the edge along its entire length, and then mark along the body with a pencil. Checking walls for square (*as shown in the photo*) is simply a matter of holding the square with the body against one wall and checking to make sure the tongue is flat against the adjacent wall. Framing squares are also used to lay out rafters and stairs.

Speed Square A speed square is a heavy-duty metal layout tool that's basically a right triangle. A lip on one edge makes it easy to use to mark stock for square or mitered cuts, *as shown in the photo at left.* Speed squares are also handy for guiding cuts with a circular saw. Just butt the saw up against the edge of the speed square and slide it over so that the blade aligns with the marked line. Holding the speed square firmly in place, make the cut.

Level and Plumb

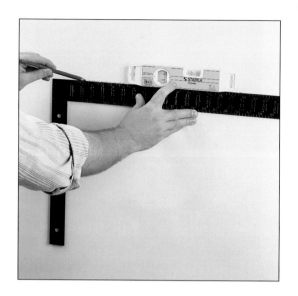

Torpedo Level Don't let the diminutive size of this tool fool you: A torpedo level is worth its weight in gold to the average framer. This small level (typically about 8" long) will shoehorn into places that your 4-foot level just can't handle. This level is so small that it easily fits into your tool pouch so that it's always on hand. By resting it on the tongue or body of a framing square, you can make quick and accurate layouts.

4-Foot Level A quality 4-foot level is an absolute necessity to the framer. Often referred to as a spirit or bubble level, a long level like this is used to check framing members for level and plumb. To use a level, hold it up against the framing member and observe the bubble inside the slightly bent fluid-filled vials. If the framing member is level or plumb, the bubble will be perfectly centered between the crosshairs of the vial. If it's not, adjust the framing member. Note: Treat your level with care and, if possible, store it in a case to prevent damage.

Plumb Bob A plumb bob is used to make sure that a framing member is perfectly vertical. It's nothing more than a pointed weight that's suspended from a string. Hold the string where you'd like to check for vertical so that the weight is just barely off the floor. Wait for the plumb bob to stop spinning or moving before you mark the point on the floor. Note: It's important to make sure that the string is centered perfectly in the plumb bob in order to get an accurate indication of plumb.

Hammer Techniques

Grip A good grip is essential for proper hammering technique. Without one, you won't transfer the maximum energy to the nail—it'll take extra blows to drive it in, which can lead to wrist pain and even tendonitis. Grip the handle with your entire hand, and wrap your thumb around the handle as far as possible (*see the photo at left*). Position the handle so the end of it is flush with the bottom of your palm *as shown*—don't "choke up" on the handle to get better control. Choking up like that would only steal the power from your stroke.

One-Handed There is a simple technique you can use when you need to start a nail one-handed. Some hammers (*like the one shown in the photo*) have a moon-shaped indentation in the back of the hammer head to accept the head of a nail. You place the nail head in the indentation, jamming the shank into the claws of the hammer. Start the nail by swinging the hammer into the workpiece. Then a flick of the wrist will disengage the nail so you can drive it home.

Face-Nail One of the two most common hammering techniques is face-nailing, where the nail is driven into the face of the workpiece, *as shown in the photo at left*. Most often this is used to secure plating, attach trimmer studs, create built-up headers, and even build walls. For nail pattern spacing for face-nailing, *see page 49*. Efficient hammering is a combination of wrist, arm, and shoulder action. Most beginners use too much wrist. Practice on scrap stock if necessary until the combined wrist/arm/shoulder motion is fluid.

Toenail The other most common hammering technique used in framing is toenailing. Here the nail is driven in at an angle to fasten two boards together (*see the photo at right*). Toenailing is necessary when you don't have access for face-nailing. Typically, you're after about a 30-degree angle here. If you're toenailing in from both sides of a framing member, stagger the position of the nails so that they don't hit each other. Note: Toenailing with an air nailer is simplicity itself; *see page 50 for more on this.*

Toenail Tip Perhaps the biggest challenge (and frustration) to toenailing is starting the nail. Often the nail will slip out of position and either fall out of the framing member or start at a bad angle. Here's an "old-timers" trick that will prevent this from happening. Simply give the stud a whack with your hammer where you want to start the nail. The face of the hammer will leave a depression in the stud that will serve as a lip or starting point for the nail. Just place the point of the nail in the corner of the depression and drive it home.

Claws As a Lifter Although not a hammering technique, using the claws of a hammer to lift and carry heavy or awkward framing members is an age-old framing tradition (*see the photo at right*). If you haven't tried this, you might be surprised how easy it makes toting a header or other part around. To use your hammer as a lifter, drive the claws firmly into the framing member with a solid swing. Then lift it up, and off you go. At your destination, simply twist and rock the hammer to disengage the claws from the wood.

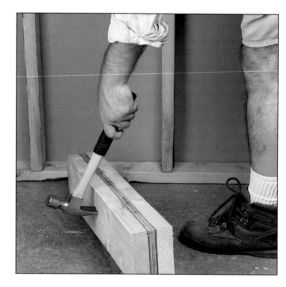

Nailing Patterns

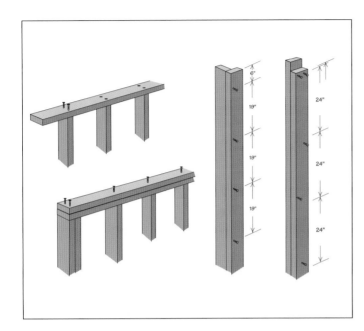

There's more to framing than just swinging a hammer: You'll also need to know what size nails to use and how far they should be spaced apart (*for more on nail types and sizes, see pages 36–37*). Fortunately, professional framers working in conjunction with home construction inspection agencies have developed nailing patterns or "schedules" over the years; *see the drawing at left.*

Please note that the patterns described here are general guidelines—you should always check your local building codes to identify accepted nailing patterns for your area.

 ## Recommended nailing schedule

Location	Nailing Technique	Nail Size and Frequency
Corner studs to blocking	Face-nail	10d – 2 on each side
Double studs: jack/king	Face-nail	10d – 16" on center, staggered and angled
Doubled 2-by header	Face-nail	10d – 2 at edges 12" on center on both sides
End stud to intersecting wall	Face-nail	16d – 12" on center to each stud
Lower top plate to stud	End-nail	16d – 2 per stud
Plywood wall sheathing	Face-nail	8d – 6" on center at edges, 12" on center at intermediate studs
Sole plate to floor	Face-nail	16d – 10" on center
Stud to sole plate	Toenail	8d – 5 per stud
	End-nail	16d – 2 per stud
Upper top plate to lower plate	Face-nail	16d – 16" on center

Air Nailers

In Use Without a doubt, the ability to toenail a stud in perfect position with the pull of a trigger is what sold me on framing nailers. Holding the gun at the proper angle, press the nosepiece firmly into the workpiece until the spiked tip grabs hold, and then continue pressing to engage the safety mechanism. Pull the trigger to drive the nail. Since a nailer drives the nail in the blink of an eye, you'll find that you don't have to have a death grip on the workpiece to hold it in position. Face-nailing with a framing nailer is simplicity itself—just press and shoot.

Stair-Stepping Stair-stepped nails, or "staircasing," is a common problem with air nailers. It's a sure sign that the compressor that's powering your nailer doesn't have enough punch. Basically, the nailer isn't getting sufficient air to drive nails to a consistent depth *as shown*. A more common example of this is where the nails stand proud of the workpiece; the first fastener will be driven all the way in, the next will be a bit proud, and the next even more. The best solution is to use a beefier compressor. If this isn't possible, slow down your nailing to let the compressor fill the tank to sufficient pressure.

STRAIGHT VS. COIL

There are two basic ways that fasteners are fed into a framing nailer: the most common *at near right,* a straight magazine (or stick), and *far right,* a coil magazine. Each has its advantages and disadvantages. A gun with a stick magazine is much lighter and far more maneuverable than one with a coil magazine. But a coil magazine offers greater fastener storage (typically 150 to 300 coils vs. 50-nail strips), which means fewer reloads.

Pulling Nails

Cat's Paw A cat's paw is one of my favorite demolition tools. That's because it frequently comes to the rescue when I need to remove a stubborn nail or one that has been driven flush with the surface of the workpiece. To use a cat's paw, position the claws directly behind the nail. Hold it at about a 30- to 45-degree angle *as shown,* and strike the head with a hammer to drive the claws under the head of the nail. Then you can pry the nail up and out. You may need to give the cat's paw a couple of solid whacks to fully drive the claws under the nail head.

Leverage Trick If a nail is proud of the work surface, you can use the claws of your hammer to pry it out. Occasionally, you'll encounter a large or stubborn nail that just doesn't want to give up its grip. In situations like this, you can either switch to a crowbar (*see below*) or use this simple trick. Just slip a scrap block of wood under the head of the hammer to provide more leverage. In many cases, this is all it takes to convince the nail to break free.

Crowbar For those extra-stubborn nails, nothing beats a large heavy-duty crowbar like the one shown *in the drawing at left.* Slip the claws of the hooked end of a crowbar around the nail head, and pull back on the bar to pull the nail free. Use your body weight whenever possible to ease the strain on your arms. Since a crowbar exerts a lot of pressure at the fulcrum point, you may want to slip a scrap of wood under the head if you need to protect the surface of the workpiece.

Using a Handsaw

Thumb Guide Although there are a number of techniques you can use to start a cut with a handsaw, the thumb method shown here is the most common. To do this, start by placing the blade of the saw on the waste side of the marked line. Then extend the thumb of your hand holding the workpiece so it extends out to touch the blade. Pull gently back three or four times to create a notch or "kerf" in the workpiece. Then, using your thumb as sort of an "outrigger" for the saw, take a full-length cut. Continue sawing, using your thumb to support the saw until you're an inch or two into the cut.

Scrap-Wood Guide If you find that you are having difficulty making a straight cut with a handsaw, consider using a block of wood to guide the saw, *as shown.* Position the block so that it's flush with the marked line. With the saw blade on the waste side of the marked line, start the cut as you did *above.* Then, keeping the side of the blade in contact with the block, saw through the workpiece. Note: If your saw isn't sharpened correctly, it can drift to one side or the other, much like a car in need of a front-end alignment. Have the saw checked by a professional sharpening service.

STANDARD VS. TOOLBOX SAWS

I like to keep two different kinds of handsaws handy for framing: a standard-length cross-cut saw and a short "toolbox" saw (*see the photo at right*). A full-length saw will make quick work of cutting framing members, since its blade allows for a longer stroke. Longer strokes mean fewer strokes. Even with the more aggressive teeth that a "toolbox" saw offers, it still takes longer to plow though a 2×6. But I find them perfect for tight spaces, and they fit in the toolbox.

Using a Circular Saw

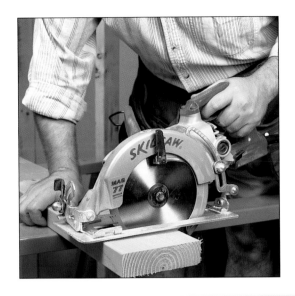

Crosscut The most common cut you'll make with a circular saw in framing work is the crosscut—cutting across the grain, such as when trimming a wall stud or plating to length; *see the photo at left* (the saw *shown here* is a "worm-drive" saw preferred by professional framers because of its power and ease of use). Place the saw on the workpiece and align the blade so it's on the waste side of the marked line. Depress the trigger and push the saw into the workpiece. Continue with one smooth motion to cut through the entire piece. **Safety Note:** Read and follow all manufacturer's directions.

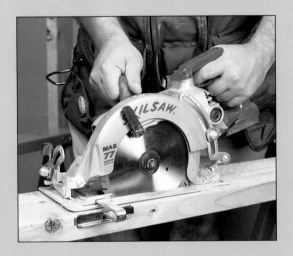

Rip Cut Occasionally, you'll need to trim or "rip" a board to width. This type of cut is made along the grain and is best accomplished with the aid of a fence (*see the photo at left*). Slide the bar into the appropriate slots in the base of the saw and adjust its position for the desired width of cut. Lock the fence in place with the clamp provided, and place the saw on the workpiece with the edge of the fence pressing against the workpiece. Depress the trigger and make the cut, taking care to keep the fence in constant contact with the edge of the workpiece as you cut.

Bevel Cut Whenever you need to cut the end or edge of a workpiece at an angle, it's referred to as a bevel cut. This is made by tilting the base plate of the saw and cutting as you would for a crosscut or rip cut (*see the photo at left*). Most saws have some type of indicator to make it easy to set the saw to the desired angle. Tilting the blade like this puts an additional strain on the saw because the blade has to remove significantly more material than if it were a 90-degree cut. With this in mind, use a slower feed rate, but keep the saw moving to prevent burn marks.

Miter Cut Although not a common framing cut, you'll use a miter cut often when applying trim to a room, especially casings around doors and windows. A miter cut is any angled cut that's made on the face of a workpiece, like that shown in the photo *at right.* Place the saw on the workpiece so the blade is on the waste side of the marked line. Depress the trigger and push the saw into the workpiece. You may find that some type of guide is helpful to make a straight cut; *see below.*

Guided A speed square is an extremely handy tool when it comes to making accurate cuts with a circular saw. Just place in on the workpiece so the lip of the square catches the edge of the workpiece. Butt the saw up against the adjacent edge of the square, and slide the saw and square so that the blade is on the waste side of the marked line. Hold the square firmly in place, depress the trigger, and push the saw through the cut, keeping the saw base in constant contact with the square.

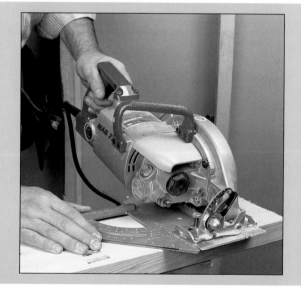

GANG CUTS

Here's a nifty trick that professional framers often use when they need to cut a number of boards to identical length. Instead of marking and cutting each piece separately, they mark one piece, hold the boards together so the ends are flush, and make a "gang" cut through all the pieces at once (*see the photo at right*). This is a great way to cut multiple cripple studs to the same length, rough sills…just about anything.

Special Cuts

Plunge Cut Square cutouts for ductwork, plumbing, and gas or electrical lines can be made by taking a "plunge" cut with a circular saw (*see the photo at left*). To make a plunge cut, position the saw so the blade is on the waste side of the cutout. Hold the front of the saw base against the workpiece and tilt the saw *as shown.* Retract the blade guard to expose the blade, and depress the trigger. Slowly lower the saw to plunge the blade into the workpiece. Move the saw forward, stopping before you reach the marked line. Do this for all four sides, and then use a handsaw to complete the cuts in the corners.

Sheet Stock Depending on your framing task, you may need to cut sheet stock for flooring or wall sheathing. The most accurate way to cut sheets is to use a shop-made or commercial straightedge like the one *shown in the photo at left.* Set the sheet stock on a pair of sawhorses, position the straightedge so the saw blade is on the waste side of the marked line, and clamp it in place. To make the cut, press the edge of the saw base firmly against the straightedge over the full length of the cut.

TRANSPORTING SHEET STOCK

A 4×8-foot sheet of plywood or OSB (oriented-strand board) is a real challenge to move around by yourself. Not only do you have the awkward size to struggle with, but also the weight. Here's a simple rope you can use to transport sheet stock. Just tie a length of rope into a loop. Then place opposite ends of the loop under the two bottom corners of the panel. Now pull up on the rope in the center of the panel to carry it (*see the photo at right*). Note: It'll take some trial and error to find the right length for the rope.

Using a Reciprocating Saw

A reciprocating saw belongs in every carpenter's or handyman's tool chest. They're terrific for making rough cuts and doing demolition work. They're indispensable for finishing cuts that other saws can't complete (such as a circular saw). Reciprocating saws accept a variety of blades that range in length from 2½" to 12". How many teeth per inch (tpi) the blade has is a good indication of its use: 3- to 10-tpi blades are wood or general-purpose blades, and 14 to 32 tpi indicates that the blade is intended for cutting metal.

The blade I use most often is a demolition blade—it's designed to cut both wood and metal, and its beefier body can handle some heavy abuse. **Shop Tip:** When you need to finish a cut near a surface, such as when cutting a sole plate near the floor, flip the blade upside down in the saw—this way you'll be able to cut almost flush with the floor.

Basic Cut To make a basic cut with a reciprocating saw, grip the rear handle firmly with one hand and then cup the body of the saw with your other hand. Position the saw on the workpiece so the blade guard is butted firmly up against the material *as shown in the photo.* This reduces vibration and gives you better control for the cut. Depress the trigger and guide the saw along the cut line. Note: On variable-speed saws, select lower speeds when cutting metal, and higher speeds for cutting wood and other materials.

Plunge Cut Although plunge cuts are possible with a reciprocating saw, they can be tricky. Place the base of the saw guard on the workpiece so that the tip of the blade isn't touching the workpiece. Depress the trigger and then, with a firm grip on the saw, slowly pivot the rear of the saw up so that the blade tip begins to make contact. It's imperative that you keep the saw extremely steady as you do this—any side-to-side movement will cause the blade to kick back. Whenever possible, drill a starter hole for the blade instead of trying a plunge cut.

Using a Saber Saw

A saber saw is anther portable power tool that comes in handy for a variety of jobs. I use mine most often when framing to make curved cutouts in sheathing for ductwork, electrical, and plumbing and gas lines. Saber saws come with either a single speed or variable speeds. A variable-speed saw allows you to match the speed to the material: high for wood, low for metal. Saber saw blades come in a huge variety of shapes, sizes, and tooth configurations. For rough work, choose a blade with 4 to 8 teeth per inch (tpi). Finer work is best done with a blade in the 10 to 20 tpi range. Blades with 24 to 32 tpi can be used to cut thin metal.

Shop Tip: Because a saber saw blade is upward-cutting—that is, the teeth cut on the up stroke—the top surface will splinter. For best results, mark your workpiece with the best side down. This way, splintering on the good side will be kept to a minimum.

Basic Cut To make a straight or curved cut with a saber saw, depress the trigger and push the saw into the workpiece. Make sure to guide the saw so that the blade cuts on the waste side of the marked line. For a tight radius cut, you'll find it easier to work around the curve if you drill a hole at the tightest point to allow the blade to pivot without binding. This technique also works well for corner cuts: Drill a ⅜" hole in each corner and cut away the majority of the waste, then go back and square up the corners with the saw.

Plunge Cut Just as with the reciprocating saw, you can make a plunge cut with a saber saw, but it's tricky. Whenever possible, drill a starter hole for the blade. If you do want to make a plunge cut, place the front of the base plate against the surface of the workpiece and tilt it forward so that the tip of the blade is not making contact with the workpiece. Then depress the trigger and pivot the saw back until the blade tip begins to cut into the surface. Once the blade breaks through, press the base plate flat against the workpiece and proceed as if it were a standard cut.

Working with Metal Framing

Although it takes a bit of getting used to, you'll find that you can put a wall up with metal framing as fast as, if not faster than, with wood. That's because you don't have to waste any time struggling with bent or warped lumber: Every metal stud is perfectly flat and straight. To further speed installation, some manufacturers "dimple" their studs to allow "snap-in" framing; *see the sidebar on page 59 for more on this.*

The biggest thing I had to get used to with metal framing is having to cut the studs to exact length—they're not forgiving like a wood stud that you can cut $1/16"$ or even $1/8"$ long and then hammer into place. If you try this with a metal stud, it'll simply kink or crumple. When this occurs, throw it in the scrap bin, as it's lost its structural integrity. Also, working with metal requires a few special tools and some extra care: A metal splinter or metal cut can be quite nasty; *see below.*

Gloves If you've never worked with metal framing before, don't worry: It's very straightforward stuff and easy to work with. One word of caution here: Whenever you're working with sheet metal, wear leather gloves. Whether it's been cut by you or by the manufacturer, you're likely to come across tiny metal slivers. These little beasties are nasty. Not only do they hurt, but they're also difficult to find and remove. Prevent this by always wearing gloves.

Cutting with Metal Snips Metal framing cuts easily with metal snips. Keep in mind that there is quite a variety of snips available—each designed for a specific cut. The cutting shears on some are optimized for cutting in a straight line, while others are optimized for curved cuts— and can either be curve-right or curve-left. When cutting framing, measure and mark all the way around it with a permanent marker. Snip the edges first, bend, and then cut through the body. Cut slowly and watch out for sharp edges and metal slivers.

Screws Metal framing members are secured to each other with sheet-metal screws. These are typically #6 or #8 self-tapping screws with a low-profile head so that they won't interfere with the installation of drywall or other wall coverings. Use a spring clamp or other clamping device to temporarily hold the metal edges together when you drive in the self-tapping screw. If you don't do this, the screw will simply push the inner stud away after it penetrates the outer stud. Make sure that the head is screwed fully against the stud to make as low a profile as possible.

Bushings Some brands of metal framing have pre-punched knockouts for running electrical lines. Because the edges of the knockouts are sharp, the steel framing manufacturers also sell plastic bushings that easily snap into place to protect the wiring. Insert a bushing into each side of the metal stud, and press the pieces together until they snap together. Note: If you're planning on using these cutouts to run wiring, take care when you cut the studs to length (and position them) so that the cutouts line up horizontally. This will simplify pulling the cable through the studs.

SNAP-IN FRAMING

Here's a nifty timesaving feature of steel framing manufactured by MarinoWare, sold under the brand name WareWall. The metal studs in this system are pre-dimpled to allow for "snap-in" framing. Dimples are located 8" on center to accommodate both 16" and 24" on-center stud spacings. To use this system, all you have to do is align the dimples on the top and bottom track, insert the stud in the tracks, and give it a twist to snap it in place (*see the drawing at right*).

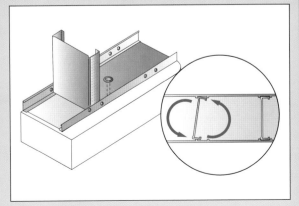

Illustration based on original artwork © MarinoWare, 2000

Straightening Lumber

Unless you're working with metal studs that are perfectly straight and flat, you're going to need to know how to straighten lumber. Even studs and framing members of the highest-quality kiln-dried lumber can twist or warp—it's the very nature of wood. In some mild cases (particularly with twist) you can persuade the lumber to behave long enough to get in a couple of fasteners; *see below.*

In the case of severe warp, you basically have two options: throw the stud out, or cut it up into smaller pieces to minimize the warp. Quite often, cutting up a stud or framing member into blocking will save it from the scrap pile. If you do have to used crooked lumber, install it where it will have the least impact (such as inner studs on a non-load-bearing partition wall). Never use crooked lumber for plating, as this serves as the foundation for a wall or wall section.

Bowed Lumber Here's a nifty trick for persuading bowed lumber (a board that's warped from end to end) to behave when you're installing two pieces face to face, such as when making a built-up header or adding a jack stud. Drive a 16d nail into the end of the straight board, *as shown in the drawing at right.* Then slip the claws of your hammer onto this nail so that the head of the hammer is positioned on the bowed board. Now pull back on the hammer so the head acts as a fulcrum to force the bowed board flush with the straight board. Have a helper drive in a few nails to hold it in place.

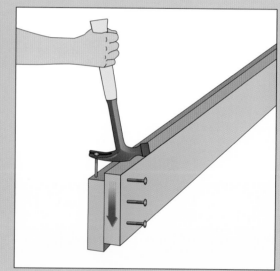

Twisted Lumber Boards that are twisted (the faces of the board aren't in the same plane along its length) can often be corrected with a shop-made "twister" (a favorite on construction sites). The twister is simply three scraps of 2-by material that are nailed (or better yet, screwed) together to form a simple fork, *as shown in the drawing at right.* The tines of the fork slip over the offending board, and the twister is bent forward or backward as necessary to remove the twist. With the twist removed, drive in nails to secure the board (you may find it easier to have a helper do this).

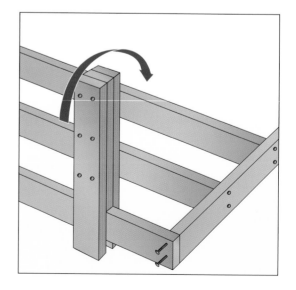

Fine-Tuning Joints

Although most framing is rough work, where studs and other framing members are often cut a bit long and then forced into place with a few blows of a hammer, other framing tasks (such as trim work) require more finesse. In a perfect world, all your molding and trim cuts would fit exactly the first time. In the real world, the standard technique is to cut the trim piece just a hair long and then "sneak" up on the final fit by trimming it to perfect length. You can use a power saw to do this, but it's extremely easy to take off too much and end up with a short piece.

Instead of this, I prefer to use two hand tools that require a little elbow grease but allow you to slowly sneak up on the perfect fit. These are the low-angle block plane and a four-in-one hand rasp (*see below*).

Block Plane One of my favorite tools for fine-tuning a joint is a low-angle block plane like the one *shown in the photo at left.* As long as the blade is sharp, you can use it to fine-tune a piece of trim or even a framing member with a few quick strokes. Make sure to hold the workpiece firmly and skew the plane to produce more of a shearing cut. Work with the grain whenever possible, and if you must plane against the grain, take precautions to prevent chip-out: Either clamp a scrap block to the edge of the wood, or plane in toward the center from both directions.

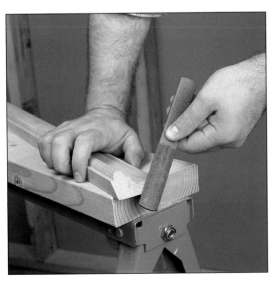

Rasp or File For really fine adjustments, such as when fitting trim or coping a joint, a rasp or fine mill file is invaluable. A four-in-one hand rasp like the one *shown in the photo at left* combines four tools in one. It features a flat and a curved rasp, and a flat and a curved file—and its small size makes it the perfect addition to a tool belt. To use a file or rasp, hold the workpiece firmly against a stable work surface like a sawhorse or workbench, and take full, smooth strokes. Most files are designed to cut in only one direction, so lift the file at the end of the cut before starting the next stroke.

Chapter 4
Framing joints

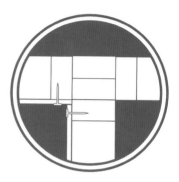

In addition to the quality of the materials you use for your framing jobs, the type of joints that you choose for connecting framing members together will have a huge impact on the strength and sturdiness of the structure. Learning the best way to attach wall studs, frame a corner, and frame a wall intersection are just some of the key building blocks you'll need to understand fully in order to take on the challenge of framing work.

Note: Although I've shown a number of common framing techniques in this chapter, it's important to check with your local building inspector to find out what the recommended framing practices are for your particular area.

In this chapter, I'll start by going over the most common framing joints: butt, edge, lap, miter, dado, and rabbet joints (*pages 63–64*). It's important to understand how each of these stacks up against the others in terms of mechanical strength and ease of use so that you can match the joint to the application.

Next, I'll describe in detail all you'll need to know to size rough openings and install them (*page 65*). Rough openings are built to house doors and windows in a wall and are designed to assume the load that the wall studs would normally bear had they not been removed to make the opening.

This is followed by the most common methods used to frame intersecting walls (*pages 66–67*)— everything from variations of blocking to a shop-made channel marker that will make laying out walls easier. Framing corners is the topic of the next section on *pages 68–69*. Two-stud, three-stud, and 2×6 corners are all described.

If the wall or wall section you're planning to build or work on has rough openings for windows or doors, you'll need to know how to make and install headers—the framing members that span the gap between the trimmer studs. *See pages 70–71* for instructions on how to use 4-by material or to build up a header from 2-by stock and plywood.

One of the best ways you can speed up framing is by accurately laying out the plating that serves as the foundation for the walls. Check out tips on how to do this, including a number of easy-to-make layout jigs on *pages 72–73*. Finally, in case you're working on a new structure, there's information on how to lay out and install floor sheathing.

Common Joints

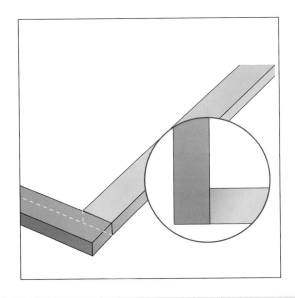

Butt Although the butt joint is the weakest of wood joints, it is the most common in construction. *As shown in the drawing at left,* a butt joint is made by simply butting two pieces of wood together. Without fasteners, this joint offers virtually no mechanical strength. But with fasteners, such as nails or screws, a butt joint provides adequate strength—and it's fast. In most cases in construction, framing members are further held in place with a covering such as plywood sheathing or drywall.

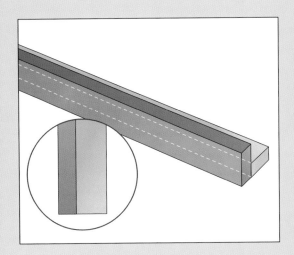

Edge The edge joint *shown in the drawing* is another version of the butt joint, where two framing members are butted up against each other. The edge joint offers a much larger surface to fasten the pieces together than the butt joint *shown above* and is regularly used in construction to frame corners and transitions where walls meet walls. *See pages 66–69* for more on corners and transitions, and *page 49* for nailing patterns for joining lumber together with an edge joint.

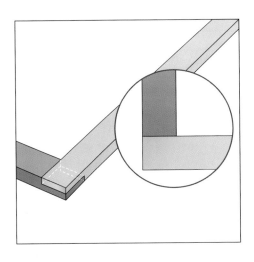

Lap A lap joint is formed whenever two framing members or trim pieces are lapped one over the other. This joint is technically referred to as a surface lap and is used often in construction for installing plating. Typically, the corner joint of a double top plate is joined with a surface lap and reinforced with nails; *see page 72 for more on this.* Another version of the lap joint is the half-lap joint, *shown in the drawing.* Rabbeting each end of the pieces to be joined helps the joint to resist lateral movement.

Miter Whenever two pieces of wood need to be joined together at a corner so that no end grain shows, a miter joint is the answer (*see the drawing at right*). Miter joints are rarely used in construction, but they are one of the most commonly used ways to join together trim pieces, such as baseboard, casing, and crown molding. Another way to join together trim pieces that leaves an almost invisible joint is the coped joint, where one of the pieces is "coped" or cut to fit over the profile of the other piece.

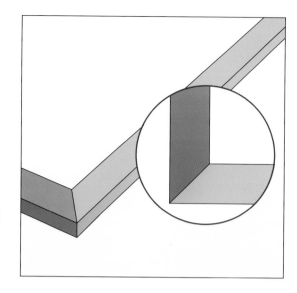

Dado A dado is a U-shaped cross-grain cut in a piece of wood that's sized to accept another part. A dado joint (*like the one shown at right*) is an excellent way to lock one part into another. Dado joints are commonly used to join together the corners of doorjambs. They're usually reinforced with nails or screws. Note: A cousin of the dado is the groove, where the U-shaped cut is made along the grain instead of across the grain.

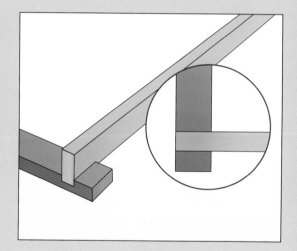

Rabbet A rabbet is an L-shaped cut made along the edge of a board either across or along the grain. The notch in the edge of the board accepts another part to form a rabbet joint (*see the drawing at right*). Like the edge joint, a rabbet joint offers a large surface area to join the pieces together. But unlike the edge joint, the rabbet serves to lock the pieces together and resist lateral movement. Like the dado joint, the rabbet joint is also commonly found connecting the corners of doorjambs.

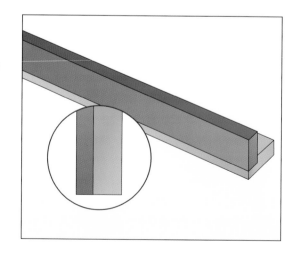

Rough Openings

Whenever you need to add a window or door to a wall, you'll need to frame a rough opening. Stud placement is critical here for the window or door to fit properly. In most cases, the rough opening should be ½" to ¾" wider and taller than the unit you're installing (consult the manufacturer's instruction sheet for the recommended gap). This extra space allows you to use shims to adjust the unit for level and plumb. In no case should you frame the opening for a wider gap. If you do, the fasteners you use to secure the unit may penetrate only into the shims and not the jack stud or trimmer stud.

Doors

The blue arrows in the door drawing *at left* identify the finished opening of the door. The green arrows show the rough opening. The framing members you'll need to install are: the king studs first, the jack or trimmer studs, and the header. Note that there's a gap at the bottom of the door for the threshold that may be installed later, or it may come as part of the unit if it's a prehung door.

Windows

Just like the drawing for the door, blue arrows *at left* indicate the finish opening and green arrows define the rough opening. The framing members for a window are similar to that of a door: king studs, jack studs, and header. The only difference is the addition of cripple studs beneath the sill plate and above the header (if the header doesn't completely fill the space above the window). Install king studs first, followed by jack studs and the header. The opening is completed by adding the cripple studs.

Double Top Plate

Header

King Stud

Door-jamb

Jack Stud

Sole Plate

Double Top Plate

Header

Jack Stud

Sill

King Stud

Cripple Stud

Sole Plate

Framing Intersecting Walls

Virtually every framing project that you take on will involve an intersection of walls. The framing for this junction needs to do two things. First, it must create a solid foundation for bracing the intersecting wall. Second, it needs to provide nailing or screwing surfaces in the inside corners for drywall or other wall coverings.

The most common method to accomplish both tasks is to add blocking, *as shown below.* The type of blocking you use will depend on the type of wall (load-bearing vs. non-load-bearing) and how tight your budget is. Blocking can be full-length, partial, 1×6, or overlapping studs. Note: Although most interior walls are framed with 2×4s, walls that carry plumbing often need to be framed with 2×6s to allow clearance for the supply and waste lines. Check your local building code to make sure which of the intersections shown here are allowed in your area.

2×4 Walls with Full Blocking The most standard method used to connect intersecting 2×4 walls is to build a U-shaped column in one wall made from three studs: two wall studs and a full-length stud referred to as blocking (*see the photo at right*). The blocking then serves as a foundation to firmly attach the end stud of the intersecting wall. This method creates full 1½"-wide surfaces in both inside corners. This supports the drywall fully and provides plenty of surface area for attaching drywall with nails or screws.

2×4 Walls with Partial Blocking If you're looking for ways to save money on a large framing project, you can use the same method as described above but with a twist. Instead of using a full-length stud as blocking, you trim cutoffs to fit between the wall studs and then face-nail blocking every 2 feet or so. This will save money on studs, but it doesn't fully support the intersecting wall as solidly as the method described *above.*

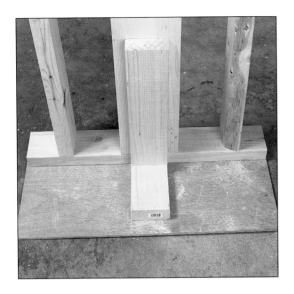

2×4 Walls with 1×6 Another alternative method to create drywall surfaces in the corners of intersecting walls is to use install a full-length 1×6 (*see the photo at left*). Since the thinner 1×6 serves as the foundation for the end stud of the intersecting wall, it's not as solid as using 2-by material. This method should be used only for non-load-bearing partition walls that don't have to support any weight. For better support, screw the 1×6 to horizontal blocking added at 2-foot intervals.

Economy 2×4 Connection The most economical way to connect intersecting 2×4 walls is to place two studs on one wall close enough together so that end stud of the inter-secting wall can be face-nailed to them (*see the photo at left*). This method creates drywall surfaces, albeit narrow ones. It also saves money since you need one less stud for each transition compared to the standard method *shown on page 66.* If you're framing a new wall or small structure like a shed, this won't amount to much; but for contractors who build large structures, the savings can be considerable.

SHOP-MADE CHANNEL MARKER

A shop-made channel marker is a quick and accurate way to lay out wall intersections and corners on plating. Just screw together two 10"-long scraps of 2×4 *as shown* so that one end of the vertical piece extends about 3" past the end of the horizontal piece.

To use the channel marker, place it at the desired location, with the vertical piece butted up firmly against the plating. Then run a pencil along each side of the horizontal piece to mark the channel.

Framing Corners

Just as there are various ways to frame intersecting walls, there are a number of methods to choose from for framing corners. Here again, the factors that affect your choice will be how strong you need the corner to be, what your budget is like, and what the local codes will allow. The three-stud method *shown below* is the most common and is widely used throughout the construction industry. Other variations use fewer studs and may or may not use blocking.

Three-Stud Corner with Blocking One of the most common ways to build a corner is to use three studs and blocking, *as shown in the photo at right.* The blocking can be full-length or partial. This type of corner is the standard for most codes, as it provides a sturdy corner and creates solid nailing or screwing surfaces for drywall. The first stud is nailed to the plating so its face is flush with the end. Then two more studs are nailed alongside this. The final stud is nailed to these so its ends are flush with the plating. Note: Although partial blocking can save material, it takes longer to install.

Three-Stud Corner without Blocking It's possible to save a wall stud at each corner location by arranging three studs *as shown in the photo.* The first stud is nailed so that its face is flush with the end of the plating. The next stud is placed along its edge and nailed in place. The last stud is nailed behind this to provide drywall surfaces on the inside corner. Note: Make sure to check your local building code to see whether this type of corner is allowed.

Two-Stud with Drywall Clips The least expensive corner you can build is also the weakest. This method uses only two studs and special metal clips commonly referred to as drywall clips or wallboard clips (*see the photo and inset at left*). Drywall clips are nailed or screwed to the stud 16" on center and have a U-shaped channel to grip the drywall. Besides its economy, this method also allows you to run insulation almost to the end of the wall. Here again, check your local building code to see whether this method of corner construction is allowed in your area. For the most part, I recommend this type of corner framing only for interior non-load-bearing partition walls.

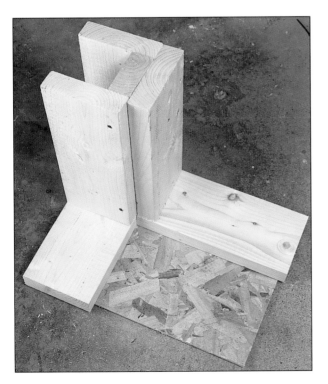

2×6 Exterior walls are often framed with 2×6s to allow additional space for insulation. Two-by-six corners can be framed *as shown in the photo at left.* This method uses three 2×6s and a single 2×4. The 2×4 is inserted as blocking between two of the 2×6s. Start by nailing the first 2×6 to the plating so that its face is flush with the end of the plate. Then position the 2×4 against its inside edge and face-nail it to the 2×6. Next, add the second 2×6 to form a U-shaped column. Finally, position the last 2×6 against the column and nail it in place.

Headers

Headers are used to span the tops of doors and windows and are sometimes referred to as lintels. They are designed to bear the weight that the wall studs that were removed (to make room for the window or door) would normally distribute. The header is supported by jack studs, often called trimmer studs or vertical trimmers. These run alongside and are fastened to a king stud. Cripple studs connect the header (and rough sill if a window is being installed) to the top plate (and sill plate for windows). Headers can be cut from 4×4 stock for a 2×4 wall or can be built up using a variety of methods (*see below*). Some local codes also allow headers to be cut from MicroLam and GlueLam (engineered beams). Local code will determine the width and length of the header, so it's important to check with your building department before planning any framing work that involves windows or doors. (*See the chart on page 71 for typical sizing.*)

4×4 Header, 2×4 Wall The most solid and quickest way to make a header for a 2×4 wall is to use a length of 4×4 (*as shown in the drawing at right*) or a 4×6. Simply measure the distance between the king studs and cut a piece of 4×4 to length. Then measure and cut cripple studs to fit between the header and the double top plate, and toenail them in place with 16d nails.

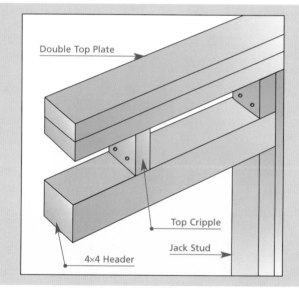

Built-Up Header, 2×4 Wall If you don't have any 4-by material on hand, you can create your own header by building it up from layers. This is usually done with two layers of 2-by material with a layer of ½" plywood sandwiched in the middle (*see the drawing at right*). Measure and cut the 2-by material to length, and then use one of these as a template to mark the plywood blocking for cutting. Trim the plywood to size and assemble the header, using 16d nails at 16" on center.

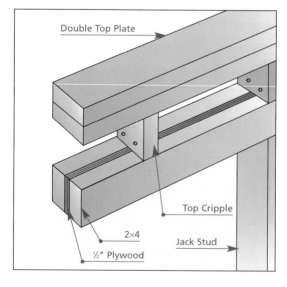

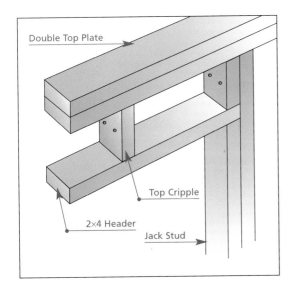

Double Top Plate

Top Cripple

2×4 Header

Jack Stud

2×4 Header for Non-Load-Bearing Wall When walls are non-load-bearing and don't have to support any weight, some local codes allow a single 2×4 header laid flat, *as shown in the drawing at left.* Measure the distance between the king studs, and cut the header to length. Face-nail it to the ends of the jack studs. Then measure and cut cripple studs to length. Position these between the header and the top plate, and toenail them in place.

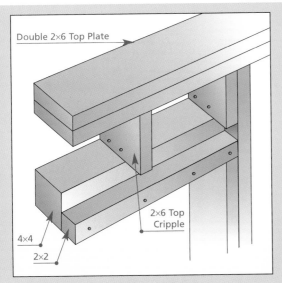

Double 2×6 Top Plate

2×6 Top Cripple

4×4

2×2

Built-Up Header, 2×6 Wall When the opening in a 2×6 wall is short, you can often get by with building up a header by using a length of 4-by material and a length of 2×2 (*see the drawing at left*). Measure and cut the 4-by material and 2×2 to length, and then nail the 2×2 to the 4-by piece. Position the built-up header on the jack studs, and toenail it in place. Then measure and cut cripple studs out of 2×6 material. Position these and then toenail them in place.

Recommended spans for headers

Size	Grade (Douglas fir)	Maximum Span (in feet)
4×4	#2	4
4×6	#2	6
4×8	#2	8
4×10	#2	10
4×12	#2	12
4×14	#1	16

Plating

Plates are the horizontal framing members that are used as the foundation to build walls—they are carefully marked to locate studs and rough openings for doors and windows, including trimmers and cripple studs. They can be either 2×4s or 2×6s. A wall typically has a single sole plate that's attached to the floor with screws or nails (walls that are attached to concrete slabs are secured with anchor bolts), and one or two top plates. Often a double top plate signifies that the wall is a load-bearing wall.

Whenever possible, run the plating the full length of the wall or wall section. If you do need to use several pieces to define a wall, try to locate the joint where the two plates butt together directly under a wall stud.

Typical Plating The drawing *at right* is an example of a couple of common plating layouts. The *left* example in the drawing is the plating layout for the corner of an interior 2×4 wall. The wall studs are *shown* as dark brown and are arranged to provide nailing or screwing surfaces for drywall on the inside corner. The *right* example in the drawing is of an exterior 2×6 wall with an intersecting 2×4 interior wall. Here again, studs are positioned to provide surfaces for drywall. (*See pages 66–67 for more framing options for intersecting walls.*)

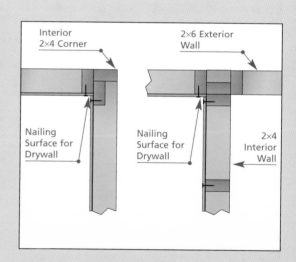

Laying Out Plating How fast a wall or structure goes up depends a lot on how much care was taken in measuring, laying out, and installing the plating. The secret to success here is to butt the sole and top plates against each other and measure and mark them at the same time, *as shown in the photo at right*. When professional framers lay out plating, they often use an "X" to define a stud, use a "T" to indicate a trimmer stud, and then usually mark any header lengths and sizes directly where the rough openings will be.

LAYOUT AIDS

There are two quick aids that you can make to speed up and increase precision of laying out plating and rough openings: a marking stick and a story pole (*see below*). Each is nothing more than a scrap or two of wood that will make your layout tasks easier.

A Marking Stick If you've got a lot of studs to mark, save yourself a lot of time by building a simple marking stick (*see the drawing at right*). The marking stick can be made out of scraps of ¼"- or ½"-thick plywood.

Cut the pieces to the sizes *shown* and fasten them together with glue and screws or nails.

To use the stick, place it on the plating so the end is flush with the end of the plating. Then make marks with a pencil on each side of the legs to locate the wall studs. Remove the stick and make an "X" at each stud location. Continue moving down the wall this way until all the studs are marked.

A Story Pole The story pole has been around in the construction industry since structures started being built. A story pole is a length of 2×4 material that's cut to equal the distance from the subfloor to the top plate.

Typically a scrap of 2×4 is fastened to the bottom to act as a sole plate. Then the location of horizontal framing members for rough openings are drawn directly on the story pole, such as headers and sill plates (*see the photo at left*). This can then be used to measure and cut the cripple studs to length.

Floor Sheathing

Regardless of the material chosen as the top layer for a floor, the underlying structure of most floors is similar. The most common type in residential construction is the framed floor. On a ground-level framed floor, the sheathing rests on joists that sit on sills along the foundation and is often supported at a midpoint by a girder. A framed floor that's elevated is typically supported by beams that run perpendicular to the joists, where the weight of the floor is borne by support columns. In most cases, the joists are tied together with bridging for extra stability and to keep them from moving side to side.

The sheathing covers the entire floor and helps to secure all the floor joists together. It serves as the foundation for building the walls. After the sheathing is down, the next step is to add the plating; *see pages 72–73 for more on this.*

Layout Floor sheathing is laid directly over the floor joists with the long edge perpendicular to the joists. Sheathing is often referred to as the subfloor or "rough" floor and is typically covered with another layer called the underlayment (usually ¼" plywood). To prevent any future problems caused by shifting materials, it's important that the underlayment be installed so that none of its seams align with any of the seams in the subfloor (*see the drawing at right*).

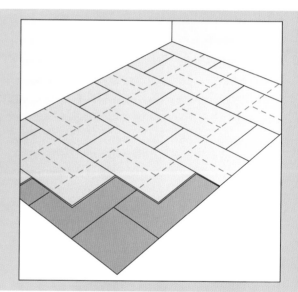

Cross Section The drawing at right is a cross section of a typical floor. Floor joists are covered with some form of subflooring, typically tongue-and-groove plywood, particleboard, or OSB (oriented-strand board). Depending on the type of flooring used, the subfloor may be covered with an additional layer of underlayment, such as cement board (for tile floors). The flooring covering is installed on top of the underlayment or subfloor and usually rests on some type of cushioning layer such as roofing felt.

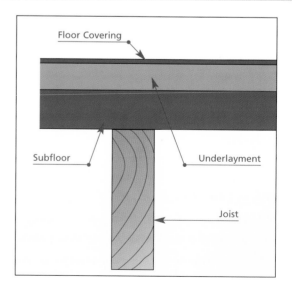

Floor Covering

Subfloor

Underlayment

Joist

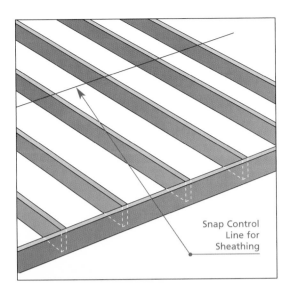

Snap Control
Line for
Sheathing

Control Line To install floor sheathing, it's best to start by snapping a control line that can then be used to align the very important first sheets. Whenever possible, start laying sheathing on the long side of the structure. To make a control line, measure in 48½" in from each end of the rim joists or room (the extra ½" allows for the tongue on the sheathing and variations in the rim joists). Align the chalk line and snap it to mark the control line (*see the drawing at left*). Use this to align the first row of sheathing.

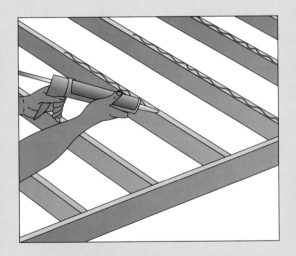

Construction Adhesive One of the best ways to prevent floor squeaks in the future is to apply construction adhesive to the tops of all the floor joists before attaching the floor sheathing (*see the drawing at left*). Apply a ¼" bead in a squiggle pattern (*like the one shown in the drawing*) to the joists. The construction adhesive will not only help prevent squeaks, but it will also strengthen the entire floor.

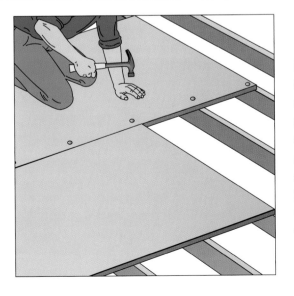

Installation With the adhesive applied to the joists, you can install the sheathing. Lay the first row of sheathing down so that the groove aligns with the control line you snapped *above* and so that each end falls over the middle of a floor joist. Then tack the sheet down in each corner and continue nailing down the sheathing, spacing nails every 6" along the perimeter of the sheet and every 12" inside the perimeter *as shown.* Take care not to damage the tongue or groove in the remaining sheets as you position them for installation.

Chapter 5
Demolition

I got my first taste of demolition work when I was 14. We had just moved into a house with a tiny kitchen. The original builder had tried to overcompensate for the small space by building a mammoth storage unit into one of the walls. It had to go, and I was the man (boy?) for the job. Armed with a pry bar, a hammer, and youthful determination, I set about the task.

Six hours and several blisters later I had learned numerous lessons from the Demolition School of Hard Knocks. First, the storage unit was site-built; that meant it wouldn't come out as a unit (like most modern kitchen cabinets) and it had to be taken apart piece by piece in the reverse order it was assembled. Since I didn't know how it was assembled, it was quite a learning experience trying to take it apart (brute force didn't work). Second, the adjacent walls were lath and plaster. I quickly learned (to my horror) that a single well-placed blow could cause cracks to appear as far as 3 feet away.

The biggest secret to demolition I've learned over the years it to do your homework up front—that is, learn as much about the area that you're planning to remove before you pick up a hammer. In this chapter, I'll start by identifying problem areas and show you what to look for

(*opposite page*). Then I'll take you through the very important steps necessary to prepare for demolition work (*pages 79–81*).

Next, we'll jump into removing a wall or section of a wall by removing the most common wall covering: drywall—everything from tips on preventing damage to adjacent walls to ripping and removing it (*pages 82–83*). For those of you with lath-and-plaster walls, I've described removal techniques on *pages 84–85*. If the wall you're removing or working on is load-bearing, you'll need to build and install temporary supports to brace the ceiling joists (*pages 86–88*). For remodeling jobs that require the removal of a door or window, see *pages 89–90* for step-by-step directions. Finally, there are detailed directions on how to remove the wall framing on *pages 91–93*.

Safety Note: Homes built prior to 1978 may have been constructed with hazardous materials: lead and asbestos. You can test painted surfaces with a test kit available at most hardware stores. Asbestos can be found in ceiling and wall materials, joint compound insulation, and flooring. Hire a professional licensed hazardous-removal company to check for this and to remove any hazardous materials that they find.

Identifying Problem Areas

You can prevent a nasty surprise when it's time for demolition work, with a little detective work up front. Granted, even the keenest house detective will still occasionally be surprised with the contents within a wall (old clothing, books, magazines—maybe even something valuable); but if you take the time before reaching for a crow bar, you'll be less surprised. Although every area of the house has its own special things to look for (*see the sections below*), the secret to identifying what's in a wall is to locate the wall from above and below (typically via the attic and basement or crawl space) and see what's going in and out of it. Basement ceilings, crawl spaces, or attics that are unfinished are a lot easier to decipher than finished areas. In either case, start by measuring from an exterior wall to the wall to be removed in the living space. Then use this measurement (referenced off the same wall) to locate the wall in question in the attic and basement. Nails protruding through the floor or ceiling indicate the location of a sole or top plate.

Basement An unfinished basement ceiling is a boon to the demolition detective. As you can see from *the photo here,* there's a lot going on—but it's easily identifiable. Although there's a jumble of electrical wiring here, none of it is going up into the floor. There are, however, two copper plumbing lines—the one on the *right* is a supply line; the larger pipe on the *left* is a waste line. The wall above to be removed is in a bathroom. Both copper lines will have to be dealt with, and there may be an old waste line in the wall, as evidenced by the empty hole in the floor in the middle of the photo.

Attic Just like a basement or crawl space, an unfinished attic will make detecting easier. In most cases, all you'll need to work around is insulation. Make sure to wear a long-sleeved shirt, gloves, and a dust mask around insulation. You'll also need to be careful moving around to stay on the joists—a pair of quarter sheets of ¾"-thick plywood will make a mobile platform that can be used to access most spaces safely. Look for ductwork and electrical, plumbing, and gas lines. *In the photo shown,* both a water heater vent and an electrical line (not to code) will come into play.

Electrical Unless you're comfortable with electrical work, moving an electrical line is best left to a licensed electrician. There are some cases where a line can be eliminated instead of moved. Switches and receptacles that have a single line coming in are "end-of-run" devices and if you desire, can be disconnected and the wiring removed back to the junction box, receptacle, or switch that feeds it power. If two sets of wires run into a switch or receptacle (*like those in the photo*), it's a "middle-of-run" device and must be moved and reconnected.

Plumbing As with electrical work, removing or moving plumbing lines should be handled by a licensed professional if you're not comfortable doing it yourself. If you're removing a line, trace it back to its source, disconnect it there, and remove the entire run. Don't be tempted to pull it below the floor or above the ceiling and just cap it; this leaves excess pipe that doesn't go anywhere and will make future improvements and repairs confusing. In most cases you don't have to go far to do this. The shower lines *in this photo* dropped down into the floor and ran over to the sink—an easy removal job.

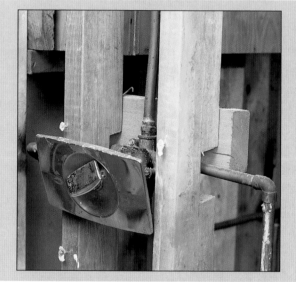

Gas Lines When it comes to moving a gas line, I strongly suggest using a professional. Although I'll tackle almost any electrical or plumbing job, I draw the line at working with gas lines. There's just too much at risk. *The photo here* shows the exterior of a cottage where the remodeling job calls for removing the window and replacing it with a door. The propane gas line (copper pipe) that runs along the siding near the bottom must be moved prior to beginning demolition work.

Preparing for Demolition

After you've identified any problem areas, you can set about getting ready for the actual demolition work. I realize it's hard to resist the temptation to pick up a sledgehammer and start in on the wall, but it's important to first prepare the area for demolition. This will save you time and trouble in the long run since you won't have to stop in the middle of the job to take care of something like turning off power or removing electrical cover plates.

If there's one other thing you can count on besides the occasional surprise inside a wall when doing demolition work, it's that you'll generate a surprising amount of dust and debris. A well-placed fan and some drop cloths will help minimize the dust problem, and a debris shoot (*see the sidebar below*) will make short work of clearing out and cleaning up the debris.

1 Locate framing members The first thing to do to prepare for demolition work is to locate the framing members in the wall or wall section to be removed. The best way to do this is with an inexpensive electronic stud finder like the one *shown in the photo.* Simply depress the ON button, press the finder on the wall, and gently slide it along. Most stud finders have visual display to indicate that you've found the edges of the stud; some finders also have an audible indicator. Mark the edges of the studs and draw an X between the lines to indicate the center of the stud.

A DEBRIS SHOOT

Depending on the amount of debris you'll be generating with your demolition work, it may be worth your while to set up a debris shoot like the one *shown in the photo.* Locate the window closest to the demolition work, remove the screen or storm window, and temporarily attach a tarp to the inside wall or window casement with duct tape. Fan out the tarp on the exterior, and simply throw or shovel debris out the window. A wheelbarrow directly under the window will make it easy to cart the debris away. Cover nearby shrubs with plywood to prevent damage.

2 **Verify stud locations** Even the most sophisticated stud finder can occasionally be fooled by a plumbing or electrical line within the wall. It's best to verify stud positions by driving a finish nail through the wall on both sides of the stud. This way you can be sure that the framing members are where you think they are. Another trick is to drill a small hole and slip in a bent wire to locate studs and/or other obstructions within the wall; *see page 119 for more on this.*

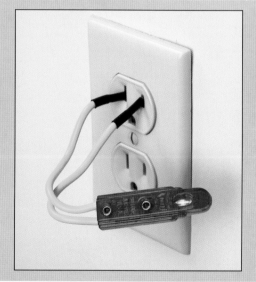

3 **Check power/shutoff** If there are any receptacles or switches in the wall or wall section to be removed, identify which circuit breaker or fuse controls their power and shut off and tag the breaker or fuse. Verify that the power is in fact off by flipping any switches to make sure the device they control is without power, and check the receptacles with a neon circuit tester, a receptacle analyzer, or a multimeter. If there's a line in the wall and you're not sure where it's going, trace it back yourself or call in an electrician to figure out what it's hooked up to—you'll have to figure it out eventually anyway.

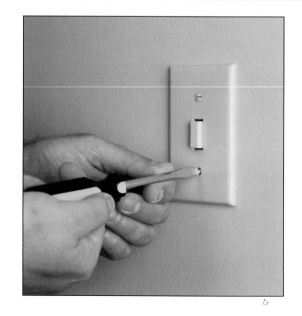

4 **Remove cover plates** Once you're sure that the power is off to all receptacles and switches, you can remove the cover plates. Use a screwdriver to loosen the screws, and lift off the plates. Don't be tempted to remove the actual receptacles or switches now; you'll be much better off waiting until the wall covering has been removed. This way it'll be simpler to identify how the wiring is set up, which in turn will make it easier to either remove or move the wiring. It's also a good idea to label any switches with a piece of masking tape to identify what they control.

5 **Drop cloths and ventilation** The next step in preparing for demolition work is to protect the surrounding area with tarps and drop cloths. Heavy-duty tarps work best to prevent nail-riddled debris from scratching floor surfaces. Lighter-weight plastic drop cloths are best used to cover furniture or other items to keep dust and debris off of them. Demolition work creates a surprising amount of dust, so it's important to wear a dust mask and also insert a fan into an open window to help exhaust the dust out of the living spaces.

6 **Remove trim** If you're planning on salvaging the trim, it's best to remove it with two putty knives and a pry bar *as shown.* Slip one against the wall, and insert a stiff-blade putty knife between it and the trim. Now insert a pry bar between the two and gently pry the trim away from the wall. This takes a bit longer, but it will prevent damage to the trim. For situations where you know you won't be reusing the existing trim, you can remove it quickly with a pry bar. With cove base molding, loosen a corner with a putty knife and then pull.

7 **Remove nails** If you're planning on reusing any of the wood trim from the wall or wall section, don't pound the nails out through the face of the molding. All that this usually accomplishes is splitting the wood, which creates a larger hole to fill once the trim is in place. Instead, pull the nails out from behind with a pair of locking pliers, *as shown in the photo.* This will create only a small hole to fill. Make sure to wear leather gloves to protect your hands from sharp nails and splinters.

Removing Drywall

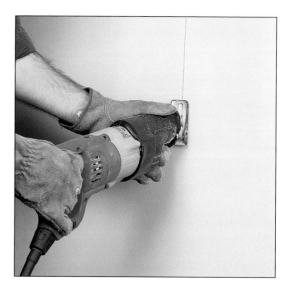

1 **Cut drywall** To remove drywall, first cut it into manageable sections so you can tear it off and dispose of it. There are a number of ways to do this. One is to set a circular saw's depth of cut to just penetrate the drywall and cut along the studs. Or you can use a reciprocating saw (*as shown*). One thing to be careful about with a reciprocating saw is that long blades can puncture hidden ductwork or electrical, gas, or plumbing lines. If you have done your detective work, you should know where these are; use a circular saw in these areas to avoid problems.

2 **Cut at joints** Like many DIYers, I'll admit that I find demolition work quite satisfying. I guess it just shows that there's a little bit of destructive energy in each of us. It is easy to get carried away—to just grab a section of drywall and rip it off the studs—but before you do this, take the time to slice through the taped joints at the corners of intersecting walls with a utility knife (*see the photo at right*). If you don't, odds are you'll rip the paper facing off the adjoining drywall surfaces and create additional work for yourself.

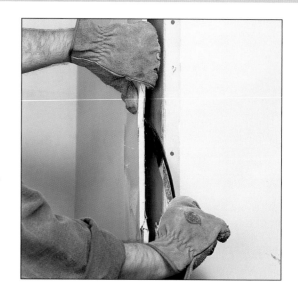

3 **Pry back and tear** Once you've cut the drywall and the adjoining tape joints, you can start to remove the drywall. The first area to work on requires the most patience and finesse—this is any section that butts up against or is next to an area of wall that you don't want to remove. Insert a pry bar under the drywall and carefully pry back the drywall until you can slip your hand in (wearing gloves, of course). Then you can carefully pull the drywall away from the studs.

4 **Hammer and pull** For sections of the wall that are not next to areas that aren't coming down, you can be a bit more aggressive with your demolition work. As long as you're sure there are no hidden surprises in the wall, use the claws of a hammer to puncture the drywall. The claws also provide an excellent purchase so you can pull and tear the wall covering off (easy there, Popeye!). With this method, a wall will quickly come down.

5 **Pull by hand** If you really want to get into it once you've got an opening, you can reach in with your gloved hands and rip the drywall off the studs (this is the part that I like). Try to pull off as large a section as possible. The larger the pieces are, the less dust you'll generate and the fewer trips you'll have to make to the debris shoot. Regardless of what method of removal you use, please remember to wear a quality dust mask (not a "nuisance" mask—they don't protect your lungs).

6 **Remove nails or screws** After all the drywall has been removed from the studs, inspect each stud carefully and remove any protruding nails with a pry bar (*see the photo at left*). Even if you're planning on removing the studs immediately, it's a good idea to do this. Protruding nails will cut though both drop cloth and tarp alike and can scratch surfaces. Nails that protrude are also likely to catch you or your clothes during removal and can easily penetrate a shoe if you step on them. For your own safety, take a few minutes to remove them.

Removing Plaster

Although removing drywall creates a mess, it's nothing compared to the mess you'll generate when removing plaster. Since plaster doesn't have a paper backing like drywall, the second you hit it, dust will fly. Also, because the wet plaster was applied directly to wood or metal lath, plenty of it was forced between the spaces in the lath to provide a purchase. This was good for the original wall but a problem now, as you'll have to pry it out from the lath. The bottom line here is that removing plaster is messy, time-consuming work. It also requires patience and a light touch. A sharp rap to a plaster surface can easily create a 2- to 3-foot-long crack. This is great if you're removing the entire wall, but bad if you want to save a portion of it. The way to get around this is to score the plaster first and then use finesse; *see Steps 1 and 2 below*. **Note:** Whenever possible, remove the entire plaster surface—it's a lot easier than trying to make a smooth transition from plaster to drywall.

1 **Score plaster near studs** Since plaster is fragile and prone to cracking, it's imperative to score the plaster section that's to be removed. Score the plaster as near to a stud as possible (*see the drawing at right*). Use a straightedge or lay down a layer or two of masking tape as a reference. To prevent cracks in the section you're not removing, you'll want to cut at least ⅛" deep, cutting in multiple passes. This takes some effort—and you'll quickly go through blades—but it's worth the time and energy.

2 **Strike to fracture** After you've scored the plaster, the next step is to strike it and break it up for removal. Tap the wall gently with a sledgehammer (*as shown*) or with the side of a hammer. The idea here is to fracture the plaster so that you can pull it away—you don't want to crush it. All that crushing would accomplish is that you would make a mess and most likely produce cracks in the wall that you wanted to save. Use gloved fingers, a pry bar, or the claws of the hammer to pry out loose pieces.

3 **Use 2×4 near edge** When you get near the line you scored in the plaster, use a hammer and a scrap piece of 2×4 to cleanly break the plaster. The 2×4 helps to distribute the blow evenly while concentrating it on the scored line. Here again, you'll want to tap gently with the hammer and take your time. Your patience here will be rewarded with much less plaster repair work. Remove loose pieces as you go, and work only on a small section at a time.

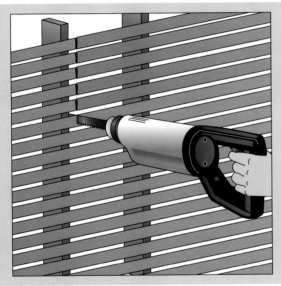

4 **Saw though the lath** Once you've got the plaster removed, you'll have to either saw through wood lath (*as shown here*) or cut through metal lath with tin snips. Cutting through wood lath is best done with a reciprocating saw fitted with a demolition blade. As always, make sure the power is off to the room and that you've identified any problem areas before cutting. If there are plumbing, gas, or electrical lines behind the lath, use a circular saw with a demolition blade to just barely cut through the lath.

5 **Pry off lath** The last step is to remove the lath itself. A pry bar will make quick work of this. When all the lath is removed, go back over the studs with the pry bar and pull out any nails. Be extremely careful how you dispose of lath. Many of the pieces will have nails protruding, and they can and will puncture a shoe if stepped on. It's a good idea to have an empty cardboard box on hand or a wheelbarrow nearby just for collecting the broken lath.

Temporary Supports

Whenever a remodeling job requires that you remove a load-bearing wall or remove more than one stud in a load-bearing wall, you'll need to make temporary supports. The temporary supports bear the weight that the wall normally would until a new support system can be installed (such as a new header or beam). The type of temporary support you use will depend on your house framing (platform or balloon) and on whether the joists run perpendicular or parallel to the wall you're working on. With platform framing, the easiest way to support the wall is to build a T-shaped support structure that can be used for either parallel or perpendicular joists (*see below*). The structure is pressed into place with hydraulic jacks (*see page 87*). Note: If you're planning on removing a load-bearing wall, you'll need to add support on each side of the wall. Balloon framing is supported by adding a temporary support header (often referred to as a whaler); *see the sidebar on page 87 for more on this.*

Platform The temporary support platform is placed under the ceiling joists roughly 3 feet from the wall to be removed or supported. As you can see *in the photo,* the support mimics a standard wall by providing a sole plate, a top plate, and support beams (in this case, beefy 4×4s). Note: It's a good idea to slip a tarp under the temporary sole plate to protect the floor from dings or dents when the jacks are used.

Anatomy To make a temporary support platform, start by measuring the rough opening you're planning on (or the width of the wall to be removed) and add 3 to 4 feet to this so the support will extend out past the opening. Cut three 2×4s to this length (or one 2×4 and a 4×4 *as shown here*). The 4×4 will be a double top plate and the single 2×4 is the sole plate. To protect the ceiling, add a layer of carpet or carpet padding to the top of the top plate. For best results, glue this in place instead of using nails or staples, since these can scratch a painted surface.

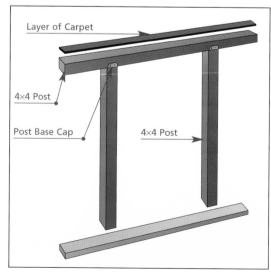

Layer of Carpet

4×4 Post

Post Base Cap

4×4 Post

Build To build the temporary support, place a pair of hydraulic supports on the sole plate and measure from the top of the jack to the ceiling. Subtract 4" for the top plate, and cut the posts to length (a post can be a 4×4 or a pair of 2×4s nailed together). Attach the posts to the top plate about 2 feet in from each end. I use metal post base caps to join the parts. Use plenty of 2"-long screws or nails to secure the base caps to the posts and top plate.

Jacks For perpendicular joists, place the jacks on the sole plate and, with the aid of a helper, lift the support onto the jacks. Adjust the posts so they're plumb, and raise the jacks up until the top plate just barely begins to raise the ceiling. Be careful—too much pressure here and you'll damage the ceiling. For joists that are parallel to the wall, bolt a pair of 4×4 cross braces to the top plate, 1 foot in from the ends, and cut posts that are 8" less than the jack-to-ceiling distance. Position the sole plate directly over a floor joist, and lift and raise the support as you would for perpendicular joists.

SUPPORTING BALLOON FRAMING

Balloon framing requires a different type of support mechanism. Instead of a T-shaped brace and hydraulic jacks, a temporary header (called a whaler) is bolted to the wall studs and supported by temporary jack studs. To support a balloon-framed wall for a new opening (such as a door or window), start by removing the wall covering from floor to ceiling. Then cut a whaler (a 2×8) about 4 feet longer than the rough opening you've planned. Then temporarily attach it to the studs so it's flush with the ceiling. Cut a pair of jack studs to fit snugly between the floor and the whaler, and attach them to the whaler with nailing plates. Next, drill holes in the whaler, and bolt the whaler to the wall studs using 3½" lag screws. Finally, drive shims under the jack studs to support the whaler firmly.

Installation How you install a set of temporary supports will depend primarily on which way the ceiling joists in your house are running with respect to the wall to which you want to remove or add a new or larger opening. If you're not sure which way your ceiling joists are running, just take a look in the attic or crawl space to determine their direction. Ceiling joists that run perpendicular to the wall in question are the easiest to support. In most cases, all that is required is a simple T-shaped cross brace and a couple of hydraulic jacks (*see page 86*).

If your ceiling joists run parallel to the wall that you want to work on or remove, it's a bit more complicated. You'll need to either provide support directly beneath them, *as shown below*, or add cross braces to the temporary supports, *as described on page 87.*

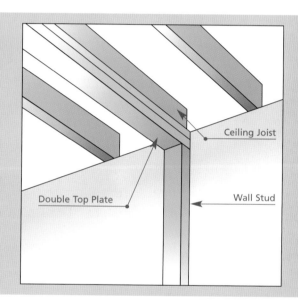

Parallel Joists Walls that run parallel to ceiling joists are for the most part non-load-bearing partition walls. If it's a non-load-bearing wall, you can modify it or remove it without temporary supports. If it is load-bearing, you'll need to use temporary supports and will have to install a new header to support the weight that the wall or wall studs previously bore. If the wall in question falls directly below the ceiling joist (*like the one shown in the drawing at right*) and it has a double top plate, it most likely is load-bearing. If in doubt, contact your local building department—they'll be able to tell you for sure.

Ceiling Joist

Double Top Plate

Wall Stud

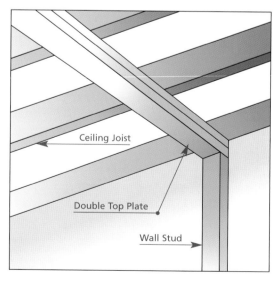

Perpendicular Joists Interior walls (*like the one shown in the drawing at right*) that have a double top plate and run perpendicular to the ceiling joists may be load-bearing walls. The best way to determine whether they are or not is to check to see whether they're aligned above a support beam. To check for this, measure from an exterior wall to the wall in question and go down into the basement and measure over from the same exterior wall. If there are support beams or columns running directly below the wall, it's load-bearing. In this case, you'll need to add temporary supports to each side of the wall before beginning work.

Ceiling Joist

Double Top Plate

Wall Stud

Removing Doors and Windows

Some remodeling jobs call for the removal of a window or door. If you're planning on simply replacing the window, follow the steps outlined *below.* If you're planning on enlarging the window opening for a larger window or a door, complete all of the interior framing work and then skip to Step 2 *below.* This way you'll minimize the time that the wall section is open and exposed to the elements. If a door or window was installed correctly, removal is pretty straightforward. If it wasn't, it can be a challenge to remove. Occasionally you'll find a window where the rough opening wasn't sized correctly and the window was forced into too small an opening. Properly sized rough openings provide clearance around the window or door so it can be shimmed level and plumb before attaching it to the framing members. With a little luck, you may be able to remove the nails that secure it, using a cat's paw (*see Step 2 below*). Otherwise, you may have to disassemble the window or door to remove it.

❶ Remove trim The first step in removing a window or door is to remove the trim. This can include casing, aprons, stools, and thresholds. In most cases, you can remove the trim with a pry bar and one or two stiff putty knives. The putty knives prevent damage to the wall and/or trim if the wall is being left intact or if the trim is being reused. To protect a wall, slip a putty knife behind the trim, insert the pry bar, and pull. If you're going to reuse the trim, insert two putty knives and insert the pry bar between them.

❷ Remove nails if possible For windows with little space between the window and the framing members, your best bet is to try to remove as many nails as possible with a cat's paw (*see the photo at left*). Be aware that a cat's paw is a destructive tool, in that it will cut into the window surface to gain a purchase on the nail. If you're planning on reusing the window, you'd be better off trying to cut through all the nails and/or screws as described on *page 90.* The only requirement for this is that you have sufficient clearance for the blade.

3 **Cut through nails** One of the most effective ways to release a window from the framing members is to cut through any nails or screws with a reciprocating saw fitted with a demolition blade, *as shown in the photo.* A couple of things here: First, you've got to have plenty of clearance for the blade—too tight an opening will result in kickback and bent blades. Second, make sure to keep the guide of the saw pressed firmly against the wall as you cut and keep a firm grip when you encounter a nail.

4 **Pry off brick molding** The next step is to remove the exterior brick molding from the framing members. Here again, a pry bar will make quick work of this. Make sure to wear leather gloves while removing the brick mold to protect your hands from nails and splinters (*see the photo at right*). If you're planning on reusing the brick mold, protect it from damage from the pry bar by slipping a stiff putty knife behind the molding as described for removing trim on *page 89.*

5 **Pull out unit** At this point the window or door should be free and ready to be removed. If it isn't, stop and find out what's holding it in place. Quite often there's a hidden nail or screw you missed. For larger windows and any size door, have a helper on hand as you pull it out of the rough opening. A pry bar may be needed to persuade a stubborn unit to come out smoothly. Note: If you're removing a double-hung window, remove the sash weights by first cutting through the sash cord and then pulling out the weights.

Removing the Framing

① **Remove surface on both sides** Before you can cut out and remove framing, you'll need to remove the wall covering from the opposite side of the wall. Since you've already got one side down, there won't be any suspense about what's inside. A few entry points will provide plenty of room for you to get your hands in and simply pull the wall covering off. As usual, pre-scoring the surface will make it easy to remove manageable-sized pieces.

② **Load- vs. non-load-bearing?** Once you've got the framing totally exposed, here's your chance to confirm that the wall is what you suspected—either load-bearing or non-load-bearing. Although the framing *shown here* might look like it's a load-bearing wall because it has a double top plate, it isn't. Both sections are non-load-bearing partition walls and could be safely removed. If you've got even the slightest doubt, call in your local building inspector. In most cases, the inspector will be able to give it a quick look and confirm which type it is.

③ **Cut studs** For load-bearing walls, make sure that you build and install temporary supports on both sides of the wall you're working on (*see pages 86–88*) before cutting any studs. A reciprocating saw will quickly zip through the studs, and it fits easily between them. If you're planning on reusing any of the studs, make your cut near either the top plate or the sole plate. If you're not, cut them in the middle, as this gives you the best "handle" to remove each piece.

4 **Remove studs** After you've cut all the wall studs to be removed, they can be taken out. Wearing leather gloves, grip each piece, bend it back toward you, and twist. In most cases, this will release the stud from the nails holding it in place. If not, lever it back and forth while twisting at the same time. Stubborn studs may need a pry bar or crowbar to convince them to give up their grip. Be careful of exposed nails in the top and sole plates. Bend over any exposed nails to reduce the chances of injury.

5 **Cut top plate** Cutting through a top plate is an awkward job regardless of whether you use a hand-saw or a power saw. I've found that a short "toolbox" saw, with its aggressive teeth, is a good choice to make the cut. Hold the saw upside down, *as shown,* and take firm strokes. Slow down as you near the completion of the cut to prevent from cutting into the ceiling. **Shop Tip:** To prevent scoring the ceiling, wrap a layer or two of duct tape around the tip of the saw—this way if it does make contact with the ceiling, it won't mar it.

6 **Remove top plate** Once you've cut through the top plate, you can remove it. The thing to watch out for here is damage to the ceiling. Since the top plate is usually secured to the ceiling joists with many nails, it won't give up its grip easily. A crowbar or pry bar usually is required to break it free. Both of these tools can cause considerable damage to your ceiling. To prevent this, insert a stiff putty knife between the bar and the ceiling and make sure to pick a leverage spot directly under a ceiling joist.

7 **Cut sole plate** The sole plate is a lot easier to cut through than a top plate, as you've got gravity working for you instead of against you. In most cases a circular saw (*as shown*) or a reciprocating saw is a good choice to make the cut. Just as with the ceiling, you'll have to take care to protect the floor from damage. The best way to do this with either of these saws is to cut only to within ⅛" or so of the floor. Then use a sharp chisel to cut through the remaining waste. On a circular saw, set the depth of cut accordingly; for a reciprocating saw, stop the cut as the blade nears the floor.

8 **Remove sole plate** The final step in removing a wall or section of a wall is to remove the sole plate. Just like the top plate, the sole plate will be firmly fixed in place. A pry bar or crowbar will most likely be necessary to lever it out. Once again, take the time to pull out or bend over any protruding nails. Nails in the floor like this are particularly hazardous because they're underfoot and, if left exposed, will surely cause an injury.

EXPOSED STUD TIP

If you're removing only a section of a wall, here's a simple tip that will save you some work. Instead of cutting the sole plate and top plate flush with the stud that will remain, leave a bit exposed for the stud or studs that you'll be attaching to the existing stud. For non-load-bearing walls, leave 1½" of sole or top plate exposed (*as shown here*). For load-bearing walls, leave 3" exposed for the double 2×4 or 4×4 post that will be installed.

Chapter 6
Adding a Wall

Adding a wall in your home is a great way to section off a portion of a large room into a special area for reading, craft, storage, or whatever. You can subdivide a basement into separate bedrooms or even add a great room. Adding non-load-bearing partition walls to your home is simple, as long as you understand the basic components of a wall and how they connect to adjoining walls.

Before you take on any of the projects in this chapter, carefully read the sections on walls (*page 13*), permits and codes (*pages 24–25*), and rough openings, wall intersections, and corners (*pages 64, 66, and 69, respectively*).

Safety Note: Adding a load-bearing wall calls for professional advice—contact your local building inspector before attempting this daunting task.

In this chapter, I'll start by showing you how to add a non-load-bearing wall by building it on the floor and then raising it up in place; *see pages 95–99.* This method is great when you've got a lot of space to work with, and it allows you to preassemble the walls. One of the big advantages to this is that you can face-nail the studs to the top and sole plates instead of toe-nailing them—face-nailing creates a stronger

joint. Next, I'll take you a step at a time through how to build a partition wall in place (*pages 99–101*). This method is best used when you have space limitations.

If you're ready to reclaim some of the wasted space in an unfinished attic, I'll show you how to add knee walls to make the space more usable (*pages 102–103*). Adding knee walls is fairly straightforward and can add much-needed living space to a home. Next, there's a section on how to reclaim living space in the basement by finishing off the masonry walls with metal studs (*pages 104–105*). Metal studs are becoming more and more popular in construction since they're inexpensive, easy to work with, and perfectly straight. But they do require some special techniques to work with; *see pages 58–59 for more on this.* Metal framing can also be used to add partition walls in a home (*pages 106–107*).

Regardless of what you use for framing members, you'll want to add a wall covering once the framing is done. The easiest way to do this is with drywall. *See pages 108–109* for detailed instructions on how to install this easy-to-work-with material.

Adding a Partition Wall

Building a partition wall is surprisingly easy. The framing is simple because partition walls are non-load-bearing and aren't designed to support any substantial weight. That's not to say you don't have to be careful—any wall you put up should be well made, firmly attached, plumb, and straight. There are a couple of options for putting up a partition wall. The method described *below* is best to use when you've got a lot of space to work in, as the wall is built on the floor and then raised into position. If your space is limited, build the wall in place (*see pages 99–101 for more on how to do this*). The method for building a partition wall described *below* uses a double top and sole plate. Although this is overkill for a non-load-bearing wall, I've found it's the easiest way to build and install a wall, as you've got plenty of clearance to lift up and install the wall. I you were to build a wall to fit perfectly between the floor and ceiling, you wouldn't have sufficient clearance to pivot it into place (another lesson from the School of Hard Knocks!).

1 **Lay out wall location** The first step in building a partition wall is to locate the nearby framing members and ceiling joists with an electronic stud finder. Then measure and lay out the wall location on the ceiling and floor. To ensure that the wall is perpendicular to the adjacent walls, use the 3-4-5 triangle method as described on *page 43*. Take your time here and double-check all measurements. Use a plumb bob (*see page 46*) to transfer the end-of-wall location from the ceiling to the floor.

2 **Cut plates** After you've located the wall, the next step is to make the top and sole plates. Carefully measure each piece separately—they quite often are not the same length (due to uneven existing walls). Transfer these measurements to four 2×4s, and cut two sets of plates to length with a circular saw, a trim saw (*as shown*), or a handsaw. Note that the top plate I'm cutting here is raised up off the floor with a pair of lumber scraps to keep from cutting into the existing hardwood floor.

3 **Mark openings** If the partition wall you're building has any openings for windows or doors, now is the time to mark them on top and sole plates. Butt the top and sole plates up against each other *as shown,* and carefully measure and mark their locations. Use a speed square (*see page 27*) to draw one set of perpendicular lines to indicate the locations of the framing members. I like to draw an "X" between the lines to make it really obvious where the studs need to go. (*See page 65 for more on rough openings.*)

4 **Mark remaining studs** With the top and sole plate still butted up against each other, measure and mark the location of each of the remaining framing members. Pay particular attention to corners and wall intersections. Here again, a speed square and an obvious "X" will help locate these. (*See pages 66–69 for options and details for corners and wall intersections.*) Note that a shop-made marking stick (*see page 73*) can also speed layout along.

5 **Attach sole plate** Once the plates have been laid out, you can attach the sole plate to the floor. Although you can use nails for this, I prefer the extra holding power of screws. Set the sole plate in place and screw down one end. Then use the 3-4-5 triangle again to double-check that it's perpendicular to the adjacent walls, and screw down the other end. Whenever possible, screw through the floor and into the floor joists. Attach a sole plate to a concrete floor using a powder-activated nailer (*see page 105*).

6 **Attach top plate** With the sole plate in place, you can turn your attention to the top plate. The first thing to do is transfer the ceiling joist locations onto it and drive screws in partway to make installation easier. Then lift it up (with the aid of a helper, if possible), butt it against the adjacent wall, and screw in one end. Check that it's aligned with the sole plate, using a plumb bob, and then screw in the other end. Then go back along its length and drive in the screws that you pre-positioned into the ceiling joists.

7 **Assemble headers if necessary** Now that the plates are in place, the next step is to build the wall. The first thing to do here is to assemble headers for any rough openings that you have planned. Headers can be assembled from two pieces of 2-by stock and a piece of ½" plywood (*as shown*). *See the chart on page 25 for recommended sizing for allowable spans.* You can also use MicroLam for spans up to 10 feet and GlueLam for up to 12-foot spans.

8 **Build wall on floor** At this point, you're ready to build the wall. Start by measuring the distance between the top plate and the sole plate, and subtract 3¼" (3" for the partition wall top plate and sole plate, and ¼" for clearance). Then position one wall stud at a time and nail it to the top plate. Continue like this until all the framing members have been secured to the top plate, and then repeat the nailing sequence for the sole plate (*see page 49 for nailing patterns*).

9 **Raise the wall** With the aid of a helper, pivot one end of the wall up and raise it until it's vertical. Then lift the wall up onto the sole plate and slide it under the top plate. Temporarily attach the new partition wall either to the adjacent wall or to the top plate with a screw or two so that you can plumb the wall (*Step 10*) and firmly secure it in place (*Step 11*).

10 **Plumb the wall** Use a 4-foot level to check the wall for plumb in both directions—at the end of the wall and on the side. Adjust the wall as necessary by inserting shims between the bottom of the wall and the sole plate. This may take a bit of trial and error in tapping shims in and out to get both the end and sides of the wall plumb. Have a helper on hand to hold the wall in place and to prevent it from sliding or creeping out of position as you do this.

11 **Secure the wall** Once the wall is plumb, secure it to the sole plate by driving nails through the shims *as shown.* Double-check for plumb before you secure the wall to the top plate. Unless you're using a framing nailer to shoot these nails in a blink of an eye, you have to be very careful that your hammering doesn't move the wall out of position as you drive in a nail. This is another instance where a well-placed screw or two can save the day—because there's no impact, the framing members have less of a tendency to shift.

Framing a Wall or Section in Place

Although I prefer to build a partition wall on the floor and then raise it up for installation (*as described on pages 95–98*), space limitations often prevent this. Here's where you'll need to build the wall in place. The major disadvantage to building a wall in place is that you'll have to toenail the wall studs to the top plate and sole plate instead of face-nailing them as when building a wall on the floor.

The problems with toenailing are twofold. First, the mechanical strength of a toenail joint isn't as strong as one that's face-nailed. Second, if you haven't spent many hours swinging a hammer, you may find toenailing a frustrating experience (*see page 48* for a nifty trick that makes this task easier). Building a wall in place also calls for added precision when measuring and cutting, since the wall studs need to fit snugly between the top plate and sole plate to facilitate nailing.

1 Lay out plates The first step in building a partition wall in place is to make and mark the top plate and sole plate. Start by measuring the length of each and then cut them to length from a piece of 2×4. Note that the wall section I'm installing in place here is a small 45-degree section of a wall (*see photo below*) that will be used to make an enclosure for a built-in storage seat. After you've cut the plates to length, butt them together *as shown* and measure and lay out intersecting walls and any wall studs.

2 Layout wall on floor The next step is to lay out the location of the wall on the floor. Take your time here, and frequently consult your plan drawing or blueprint to make sure you're transferring measurements correctly. Use the 3-4-5 triangle method as described on *page 43* to make sure the new wall or walls will be perpendicular to existing adjoining walls.

3 **Extend wall location to ceiling** Once you've marked the wall location on the floor, you can transfer its location to the ceiling with a plumb bob so that you can locate the top plate. Hold the string of the plumb bob up against the ceiling so that the plumb bob hovers directly over each corner of the sole plate—you'll need a helper to steady the plumb bob and make sure it's in the correct spot before you mark the ceiling. If you take your time here, you'll end up with a series of dots that can be connected to define the top plate.

4 **Lay out top plate** Use a framing square or other straightedge to connect the dots you made in Step 3. Alternatively, you can measure and mark from a single dot location as long as you're using a square and are measuring carefully. Even if you connect the dots, you should go back over the outline and measure it to make sure it's the correct width and length. Here again, it's worth the time to use the 3-4-5 triangle to check for perpendicular.

5 **Secure top plate to ceiling joist** After you've laid out the top plate or plates, you can attach them to the ceiling. The first thing to do is mark the ceiling joist locations on the top plates and drive in a screw partway at each location. This makes it a lot easier to position the plate and secure it. Position the top plate and screw one end to a ceiling joist. Then check for perpendicular one more time and screw in the other end. Then go back over the length of the plate and drive any remaining screws into ceiling joists.

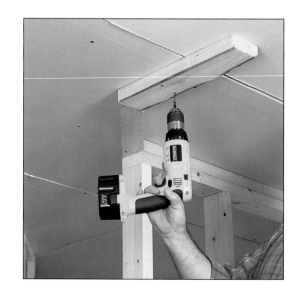

6 **Attach sole plate** Now you can attach the sole plate or sole plates to the floor. Butt one end up against the existing wall and drive in a screw. Check the plate for perpendicular, and drive a screw in the other end. Continue working along the plate, driving in any remaining screws. Note: For hardwood floors like the one shown, you may find it easier to drive in screws if you first drill a pilot hole. You can use a powder-activated nailer to attach sole plates; *see page 105 for more on this.*

7 **Toenail the studs** With the top plate and sole plate in place, you can add the wall studs. Measure and cut one stud at a time, because odds are that the ceiling and floor are not parallel to each other. Cut the studs to fit snugly between the plates—a tight fit helps hold the stud in place for nailing. Toenail each of the studs to the top plate and sole plate (*for more on toenailing, see page 48*). If you've got access to a framing nailer, you'll really appreciate how easily they handle this often frustrating task (*see page 50 for more about air-powered nailers*).

8 **Add additional framing members** The final framing step is to add any framing members for intersecting walls, corners (*as shown here*), or rough openings. Here again, it's best to measure, cut, and install one piece at a time to compensate for uneven walls, floors, and ceilings. When the framing is complete, have your local building inspector check your work before applying a wall covering (*see pages 108–109 for directions on how to install drywall*).

A Knee Wall in the Attic

Building a knee wall in the attic is a fairly straightforward and simple job. The toughest part is deciding on the distance "A" *in the drawings below*—how far out the knee wall extends into the attic. As it extends into the attic, square footage drops, but wall height increases. I'd advise mocking up the walls with cardboard or foam board, in the planning stages, to get a solid feel for how this will affect the usability of the room. Many homeowners simply install a door or two in the knee wall and use the space behind the wall for storage.

The only other challenge involved is bevel-ripping the top plate and then cutting the wall studs to match the angle. But a plumb bob will identify the top plate angle, and there's a simple trick to obtain the angle for the wall studs (*see Step 3 on page 103*).

Knee wall anatomy A knee wall consists of four main parts: a sole plate that's attached to the attic floor, a beveled top plate that's screwed to the rafters, backer boards that supply screwing or nailing surfaces for drywall, and wall studs that connect the sole plate to the top plate; *see the drawing at right.* Although you can install a top plate without beveling it by setting it back so the wall studs are flush with the front corner, this method doesn't offer any drywall surface along the top edge.

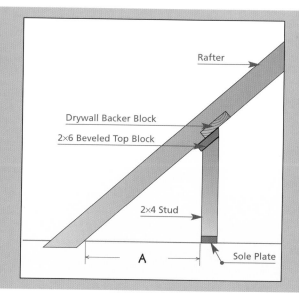

1 **Attach sole plate** Once you've decided on the placement of the knee wall, measure out from the roof the desired distance on both ends of the area you're working on and snap a chalk line as a reference for the sole plate. Cut the sole plates to length, position them along the chalk line, and attach them to the attic floor with either nails or screws. Drive an extra nail or screw in wherever the sole plate passes directly over a ceiling joist to secure it firmly.

2 **Attach top plate** Once you've installed the sole plate, use a plumb bob to transfer the wall location to the rafters at both ends of the wall (and along the wall as necessary). Bevel-rip the top plates from 2×6 stock (*see page 53 for instructions on how to do this*), and then measure and cut them to length. Align each top plate with the marks on the rafters and secure them with nails or screws (a helper will make this work a lot easier).

3 **Install studs** The next step is to install the studs. Measure and cut these one at a time to compensate for uneven floors and rafters. **Shop Tip:** To find the correct angle to cut the tops of the studs, cut a scrap of 2×4 to fit between the floor and the roof—hold it against the sole and top plate and butt it against a rafter. Then use a pencil to scribe the rafter angle onto the scrap block. Cut the scrap to this angle and then use it to transfer the angle onto the wall studs. Cut this angle first, then trim the stud to the desired length. Toenail each stud in place.

Backer
Block

4 **Add backer blocks** The final framing step is to add drywall backer blocks between the rafter to provide a nailing or screwing surface for drywall. Measure and cut these one at a time because the distance between each rafter will most likely be different. Start at one end of the wall and face-nail each backer block to the appropriate rafters, taking care to make sure the face of the block is flush with the edge of the rafters. Once they're in place, you can add the wall covering of your choice (*see pages 108–109 for instructions on installing drywall*).

Finishing a Wall with Metal Studs

Exposed masonry walls are a challenge to finish off for two main reasons: irregularities and moisture. In many homes, the concrete, cinder block, stone, or other masonry walls in the basement are hugely irregular in terms of how flat, straight, and plumb they are (or aren't). The simplest way around this is to build new partition walls a set distance away from the masonry. Although this reduces the overall square footage of the area, it does provide for flat, plumb walls.

The moisture problem, on the other hand, isn't so easily handled. If you've got moisture problems in the basement, you'll need to have them repaired before you add new walls. If you don't, the moisture seeping in will quickly ruin all your work. If you don't have an obvious moisture problem, you'll still need to install a vapor barrier because any masonry surface acts as a moisture conduit to some extent (*see Step 1 below*).

1 **Install vapor barrier** Check your masonry wall for moisture problems and have any repaired before you add new walls. The first thing you'll want to do is to add a vapor barrier to prevent any moisture problems with the wall covering (all masonry surfaces, even those in good shape, will still allow some moisture to seep through). Cut 6-mil plastic vapor barrier to size, and attach it directly to the masonry with mastic or construction adhesive. Overlap the seams 6", and consider adding vertical furring strips to help hold the vapor barrier in place over time.

2 **Bottom track** Measure out from the wall about ½" or so for air circulation and to compensate for wall irregularities. Strike a chalk line along the floor for reference, and measure and cut a bottom track to fit (*see pages 58–59 for more on working with metal framing*). Position the bottom track along the chalk line you snapped, and secure the track to the concrete floor using appropriate fasteners and a powder-activated nailer (*see the sidebar on the opposite page*).

3 **Top track** Use a plumb bob to transfer the location of the bottom track to the ceiling joists to locate the top track. (In the basement *shown here,* the wall we're adding runs parallel to the ceiling joists.) Once you've marked the location of the top track, secure it to the ceiling joists with self-tapping screws or nails (*see page 59 for more on self-tapping screws*).

4 **Studs** The next step is to add the wall studs and rough opening framing members. Measure and cut each stud one at a time to compensate for uneven floors and ceilings. Position the wall stud, check it for plumb, and secure it with sheet-metal screws. Once all the wall studs are in place, cut and measure framing members for rough openings (if any) and install them as you would for the wall studs. Windows (*like the one shown here*) are easily framed with metal studs.

POWDER-ACTIVATED NAILER

The easiest way to attach framing to concrete is to use a powder-activated nailer (often called a power hammer). A powder-activated nailer is basically a gun barrel that's activated by a trigger or by striking it with a hammer. This discharges a hardened steel fastener expelled by a bulletlike power load (actually a .22 to .26 blank cartridge). For this system to work, you must use the appropriate size load and fastener. Charts on the nailer and fastener packaging will help you determine the correct size based on what you're driving the fastener into and the length of the fastener. Powder-activated nailers are both loud and dangerous—wear ear and eye protection, and follow the manufacturer's directions.

Adding a Wall with Metal Studs

Adding a partition wall with metal framing is very similar to installing a wall using wood studs. The basic construction concepts are the same: You've got a top plate and sole plate (top and bottom track) that are connected with studs. The big difference is how you cut and attach the framing members. Working with metal studs isn't for everybody. If you're used to cutting a stud a bit long and then pounding it into submission with a hammer, think again. This just doesn't work with a metal stud—any force can cause it to buckle. Also trimming just a little bit off the end of a stud is a hassle, since most metal snips work best with at least ¼" of material on each side of the cutters. On the plus side, if you use the snap-in type studs *as shown here,* the walls go up surprisingly fast. Just remember, you'll need special self-tapping screws to work with this stuff (*for more on working with metal framing, see pages 58–59*).

1 **Top track** The first step to installing a metal partition wall is to lay out the top track. Measure out from a nearby wall, and snap a chalk line for reference. Use the 3-4-5 triangle method to make sure the new wall will be perpendicular to the adjoining walls. Measure and cut the top track to length with metal snips. Hold the track so it's aligned with the chalk line, and drive in a screw at one end through the track and into the ceiling joist. Check for perpendicular one more time, and secure the other end to a ceiling joist. When possible, work along the length of the track, screwing it to the ceiling joist.

MOLLY BOLTS

Even with careful planning, you'll occasionally need to install a top or bottom track or stud to a connecting wall where there aren't as many ceiling joists or studs available to screw into as you'd prefer. In cases like this, and only when you're installing a non-load-bearing partition wall, you can use molly bolts to help secure the framing member to the wall or ceiling (in addition to the framing members into which you can drive fasteners). The metal wings of a Molly bolt (also known as an expansion bolt) expand out as the screw is tightened and lock the fastener to the wall. Molly bolts are sized to match the thickness of your drywall.

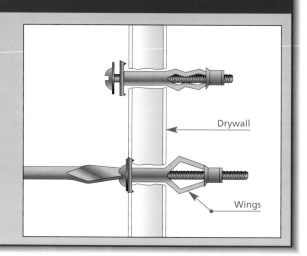

Drywall

Wings

2 **Plumb** With the top track in place, the next step is to locate the bottom track for the wall. Suspend a plumb bob from the top track and have a helper transfer this location onto the floor. Do this at each corner of the top track, and you should be able to mark the location of the bottom track by connecting the dots. Use the 3-4-5 triangle to verify that the bottom track will be perpendicular to the adjoining walls.

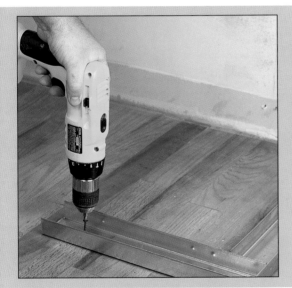

3 **Bottom track** Measure and cut a bottom track or tracks to length and set them in position on the floor. Butt one end of the track against the adjoining wall, and carefully align the track with the layout lines on the floor. Secure one end of the track to the floor with self-tapping screws. Then check again for perpendicular and secure the other end of the track. Work along the track, fastening it to the floor every 8" or so with self-tapping screws. With hardwood floors, *like the one shown,* it's best to drill pilot holes before driving in the screws.

4 **Studs** All that's left for framing is to add the wall studs. Measure and cut each stud independently to compensate for uneven floors and ceilings. Position the stud and secure it to the top and bottom tracks with self-tapping screws. When all the studs are in place, add any framing members for rough openings. Have the local building inspector check your work, and then install the wall covering of your choice (*see pages 108–109 for instructions on installing drywall*).

Installing Drywall

1 **Put up first sheet** Once the framing is up, you can add a wall covering (make sure to have your work inspected before doing this). Unless you're doing renovation work in an older house where you want to match the existing plaster, the easiest wall covering to install is drywall. I recommend ½" drywall, as it holds up better over time. Start by positioning the first sheet tight in the corner, and screw it in place. (Note: Professionals often install sheets horizontally, *as shown here,* because this makes taping easier.) Drive drywall screws in so they sit just below the surface, but don't break through the paper covering.

2 **Cut to fit** The second sheet will most likely need to be cut to fit. Carefully measure from the existing sheet to the ceiling or floor on both ends of the panel and transfer these measurements to a full sheet. Draw a line with a straightedge or make a chalk line, and then cut along this line with a sharp utility knife. Flip the sheet over and lift up one end to snap the sheet. Run your utility knife along the inside crease to cut completely through the sheet. Check the fit of the cut sheet, and trim as necessary. Attach the sheet with drywall screws, and continue until the framing is covered.

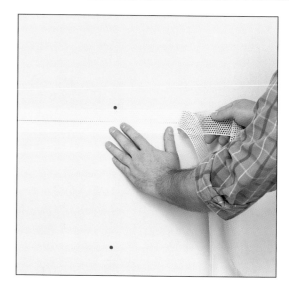

3 **Tape the joints** To conceal the joints between the sheets of drywall, apply drywall tape over the gaps. Drywall tape may be self-adhesive (*as shown here*) or nonstick. To apply self-adhesive tape, simply remove the paper backing and press it in place. Nonadhesive tape is applied by first spreading on a thin coat of joint compound and then pressing the tape into the compound with a wide-blade putty knife. Exposed corners are best covered with metal corner bead—thin L-shaped metal that's easily cut with tin snips and is then either nailed or screwed to the framing members.

④ Apply first coat With all the tape in place, the next step is to apply a first coat of joint compound. Joint compound comes premixed or in a powder form that you can mix yourself. Apply a generous first coat with a 4"- to 6"-wide drywall knife or putty knife. Cover the tape completely and also all the impressions or "dimples" left by the screws. Inside and outside corners are best done with special drywall tools designed especially for this. Apply the joint compound as smoothly as possible, but don't be too meticulous—you'll remove any high spots before applying the next coat.

⑤ Feather Once the joint compound you applied in Step 4 has dried thoroughly (usually overnight), go over the joints with a stiff-blade putty knife and knock off or scrape away any high spots. Then apply the next coat. Use a wider drywall knife, and spread the joint compound over the first coat. Work the compound gently away from the joint to "feather" it for a smooth transition to the drywall. Multiple thin coats work best here. Let each coat dry, knock off any high spots, and then apply the next. Repeat as necessary to create a flat surface.

⑥ Sponge When the joint compound is completely dry, the final step is to smooth the surface to remove any remaining imperfections. In the past, this was usually done with sandpaper or sanding screen, and it created a horrendous mess. A tidier alternative that works great is to smooth the joint compound with a drywall sponge like the one shown here. These sponges have an abrasive pad glued to one side to quickly flatten high spots. Wet the sponge and wring it out so it's just damp. Then use a swirling motion to smooth the joints.

Chapter 7
Adding Built-Ins

Without a doubt, built-ins are one of the most personal ways you can customize your house. Built-ins offer exciting ways to add storage space, decorative touches, and even seating to any living space in your home. If this is true, why are there so few built-ins in most homes? Well, there are a couple reasons for this. First, builders tend to steer clear of these because they are custom work that can add significantly to the price of a home. Second, since most built-in work by its very nature is site-built custom work, hiring it out can be expensive. And third, because most built-ins require knocking out a portion of a wall, most homeowners are wary of the task.

But adding a built-in to your home isn't so difficult. All it takes is some work up front to identify a likely spot for a built-in and then a little detective work to make sure there won't be any surprises when you open up a wall. Although most built-ins have to be made at the site, some (like medicine cabinets and some shelving units) come premade; all you have to do to install one of these is create the appropriate-sized opening in the wall. Site-built versions are easy to make with readily available materials, a few tools, and some simple joinery.

In this chapter, I'll start by going over the anatomy of a built-in and identifying the terms used to describe them (*opposite page*). Then I'll take you a step at a time through how to add four commonly requested built-ins: built-in seating, adding a closet, built-in wall storage, and adding a built-in medicine cabinet. The built-in seating unit described on *pages 112–115* features plenty of storage underneath that's accessible by way of a hinged seat. The one shown there has a sturdy 2×4 frame and is covered with oak plywood and trimmed out with matching oak molding.

I don't know anyone who wouldn't like a little more closet space in some room in their house. The built-in closet shown on *pages 116–118* is made with metal framing and covered in drywall. Another storage project is the built-in wall storage unit shown on *pages 119–122*. It features glass shelves and recessed lighting. Finally, the medicine cabinet shown on *pages 123–125* has built-in lighting and mirror doors. Although the directions are specific to the projects shown, the general techniques can be used to add almost any built-in to your home.

Built-In Anatomy

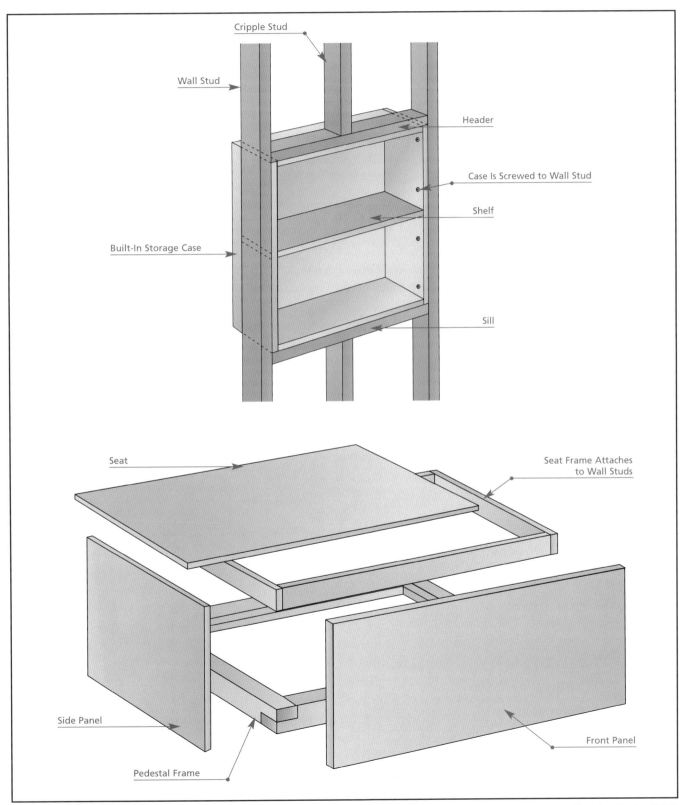

Cripple Stud

Wall Stud

Header

Case Is Screwed to Wall Stud

Shelf

Built-In Storage Case

Sill

Seat

Seat Frame Attaches to Wall Studs

Side Panel

Pedestal Frame

Front Panel

Built-In Seating

1 **Measure and cut wall cleats** The first step in adding built-in seating is to measure and cut the wall cleat that will support the weight of the back of the seat. The wall cleat runs the length of the wall and is firmly screwed or bolted to the wall studs. Measure the distance from wall to wall with a tape measure *as shown.* Then cut a 2×4 wall cleat to this length and check to make sure that it fits snug but not too tight in the space. Next, measure and cut two shorter wall cleats for the adjacent wall.

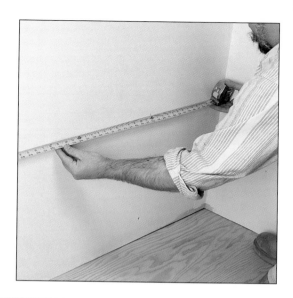

2 **Attach wall cleats** To attach the wall cleats, start by locating the wall studs, using an electronic stud finder. Then measure up from the floor the desired height (typically, 18") and make a mark on all three walls. Since most floors aren't perfectly level, position the cleats using a level before screwing or bolting them in place (*as shown*). Make sure to keep the cleat aligned with the mark you made in Step 1 as you do this. Three-inch-long deck screws work well, or for added security use 3" lag screws to attach the cleats to the wall studs.

3 **Locate sole plate** With the wall cleats in place, the next step is to transfer the width of the seat to the floor so that you can install a sole plate. Normally you'd use a plumb bob for this; but since the wall cleats are so close to the floor, a framing square will do the job. Butt the long blade of the square up against the end of the wall cleat *as shown,* and make a mark on the floor to indicate the sole plate's position. Note: If you haven't already removed the section of the baseboard under the seat, now's the time for this (just make sure to take into account the thickness of the front panel).

4 **Attach sole plate** Once you've located both ends of the sole plate using the framing square, measure the wall-to-wall distance and cut a piece of 2×4 to length. Then position the sole plate on your marks and screw it to the floor. Note that with some types of flooring (like the hardwood floor *shown here*), you may need to drill pilot holes before driving in the screws. Nails will also work here, but I prefer the better holding power that screws offer over nails.

5 **Add frame pieces** All that's left to complete the frame is to add the frame support pieces: the front rail that spans the wall cleats, and three short support pieces that fit between the sole plate and front rail. To do this, start by measuring the front rail, cut it to length, and screw it to the ends of the wall cleats. Then measure the distance between the sole plate and the front rail and cut three supports to fit. Screw the end supports to the wall as well as to the sole plate and front rail. Screw the middle support to the sole plate and front rail *as shown*.

6 **Attach plywood front** Now that the frame is complete, you can add the plywood front and top. Start by carefully measuring the wall-to-wall distance and the distance from the bottom of the sole plate to the top of the front rail. Be sure to take measurements at both ends of the seat to compensate for uneven walls and floors. Transfer these measurements to a sheet of ¾"-thick hardwood plywood (I used red oak) and, using a straightedge and a circular saw, cut it to size. Attach the front to the frame with finish nails (I used an air-powered finish nailer for this).

7 **Make the seat** The seat is a rectangular piece of plywood that's cut to fit between the walls so that it ends up flush with the front rail. If you want to hinge the seat to use the space below for storage, subtract 1" from the width of the seat to allow for a hinge and a hinge cleat that's attached to the wall (*see Step 8*). One of the walls that the seat shown here fits between is at a 45-degree angle. Instead of hinging the entire seat, I chose to hinge a 4-foot section (a common hinge length) and then cut and permanently install a 45-degree piece to the seat frame.

8 **Attach piano hinge** The best way to fully support the weight of the seat and allow it to swing up to access the storage space below is to use a piano hinge, *like the one shown here.* To install a piano hinge, start by cutting a ¾"-square cleat to fit between the walls. Then attach the hinge to the cleat and then to the seat (a self-centering drill bit or "Vix" bit comes in really handy here). Then you can attach the hinge cleat to the wall by screwing it to the same wall studs that you used to attach the wall cleats.

PREFINISHED PLYWOOD

If you're ever faced with a large project that calls for a lot of exposed plywood, such as a built-in bookcase, built-in seating, or wainscoting, you might want to consider using prefinished plywood. Prefinished plywood has been used in the high-end cabinet-making industry for years, but it's popping up more and more in smaller shops where time-saving benefits can easily outweigh the added cost of the plywood. A wide variety of natural and stained or dyed hardwoods are available, most with UV-cured finishes. Contact States Industries at www.statesind.com for more information, for your nearest retailer, and for pricing.

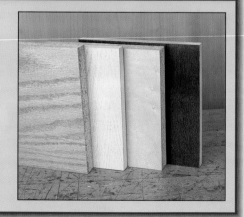

9 **Add lip to seat front** To cover the exposed edge of the plywood seat and add a nice decorative touch to the top of the seat front, I attached a lip to the edge of the seat. The lip *shown here* is a piece of oak molding that's ⅜" thick and 1" wide. Cut the lip to match the length of the seat, and attach it with glue and brads (I used an air-powered brad nailer). For added strength, you can drill counterbored holes, screw the lip to the seat, and then conceal the holes with wood plugs.

10 **Add trim** The last pieces to add to the built-in seat are the trim. Depending on the look you're after, you can add matching hardwood trim (*as shown here*) or you can paint the trim to match the rest of the trim in the room. You'll need base trim and a couple of short pieces of quarter-round molding to conceal any gaps between the ends of the plywood front and the adjacent walls. Attach the trim with finish nails or brads, and then use a nail set to countersink the nail heads below the surface (do this for all exposed nails).

11 **Apply finish** Before you apply the finish of your choice, fill any nail holes with wood putty (for a natural finish) or painter's putty (if you'll be painting the seat). When the putty is dry, carefully sand all the parts of the seat with 120-grit sandpaper. Then vacuum away all the sanding dust, mask off adjacent areas with masking tape, and apply a finish. For a natural finish (*as shown here*), I recommend two coats of satin polyurethane, sanding lightly between coats with 220-grit wet/dry sandpaper. If you're planning on painting the seat, apply a quality primer before painting.

Adding a Closet

1 **Lay out top plate** Adding a closet is really quite
simple if you do your homework first. Start by locating wall studs and ceiling joists in the area where you
want the closet. Whenever possible, size the closet so that
you can attach the wall directly to existing wall studs and
ceiling joists. (The standard depth of a closet for hanging
clothes is 20" to 24".) Measure out on the ceiling from the
wall opposite the intended doorway of the closet to
locate the top plate. Do this at both ends of the closet.

2 **Attach top plate** Once you've located the top
plate, you can cut it to length and attach it to the
ceiling joists. I used metal framing here (*see pages 58–59
for more on this*), but you could just as easily use 2×4s. If
you're planning on using metal framing, you'll need to
purchase self-tapping screws (*top in inset*). These screws
have a finer thread than drywall screws (*bottom in
inset*) and have a pointed tip that will drill easily
through the metal framing. Make sure that each
screw is firmly attached to a joist.

3 **Plumb bob to floor** Now you can transfer the position of the top plate to the floor to locate the sole
plate. Don't be tempted to simply measure out from the
wall to do this. Odds are that the walls are not plumb, so
if you do this, the walls of your closet won't be plumb,
either. Instead, use a plumb bob *as shown here.* Hold the
string of the plumb bob against the outer edge of the top
plate, and mark the corresponding position on the floor—
do this for both ends of the top plate.

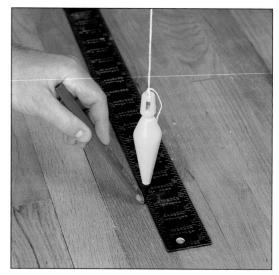

4 **Attach sole plates** After you've located the sole plates, you can measure and cut the pieces to length. You'll have to take into account the size of the rough opening for the door when you do this (*see page 65 for more on rough openings*). Align each sole plate piece with the marks you made in Step 3, and screw each piece to the floor. Note: If you're screwing into a tough floor surface (like the hardwood flooring *shown here*), it's best to drill pilot holes first before driving in the screws.

5 **Screw studs to wall** With both the top and sole plates in position, the next step is to attach studs to the adjoining walls. Since the closet walls are partition walls and are not load-bearing, you can screw them directly to the adjoining walls. If you were able to size the closet to hit the wall studs, this is simply a matter of cutting the studs to length and then screwing them in place. If not, use Molly bolts (*see page 106*) to anchor them firmly in place.

6 **Attach studs** Now you can cut the remaining full-length studs to length, position them, and secure them to the top plate and sole plate with screws (*see page 59 for more on this*). Drive the sheet-metal screws in far enough that they're flush with the framing member and don't protrude. Also, make sure that you build the corners so that you're providing drywall surfaces in the interior of the closet (*see pages 68–69 for more on framing corners*).

7 **Attach headers** All that's left for framing is to measure and cut the jack studs and headers to length for the rough opening for the door. Start by cutting the jack studs to length, and attach them to the sole plate and the king studs. Then cut the header pieces to fit between the king studs and secure them to the king studs and jack studs with screws. Note: Since this is a non-load-bearing partition wall, you can get by with a single 2×4 header, as long as you install cripple studs between the header and the top plate.

8 **Install drywall** The closet will really begin to take shape as you install the drywall or wall covering of your choice. Here again, if you're using metal studs (*like those shown here*), you'll need to use self-tapping screws to secure the drywall to the studs. Attach the drywall to the studs every 8" or so, and then continue to work around the closet until all the drywall is in place.

9 **Add door casing** To complete the closet, add the casing to the rough opening for the door. Measure and cut the pieces to length following the manufacturer's directions. Then attach the top casing piece to the side pieces and position this unit in the rough opening. Add carpenter's shims as needed to plumb the casing—check this with a 4-foot level *as shown*. When the casing is plumb, secure it to the framing members by driving screws through the casing and shims. Snap off the excess shims, and finish the drywall as described on *pages 108–109*.

Built-In Storage

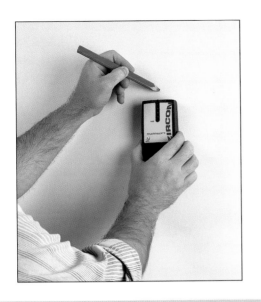

1 **Locate studs** The secret to successfully installing a built-in wall storage unit is identifying a location that has sufficient clearance behind the wall for the unit. If all you want is a narrow display shelf, you can build one to fit in any 2×4 wall that doesn't contain any ductwork or plumbing or any electrical or gas lines. A bit of detective work (*see pages 77–78*) will identify most of these. If you've got a set of updated blueprints for your house, you can use these to locate prime storage spots. Once you've identified a wall, locate the wall studs with a stud finder.

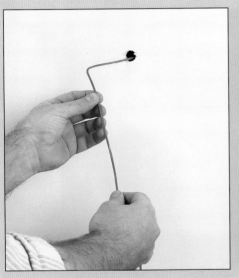

2 **Drill exploratory hole** Even with the most thorough detective work, you can occasionally miss a clue and discover a surprise when you open up the wall. To prevent this from happening, I always drill an exploratory hole in the wall centered between the wall studs with a ¾" spade bit. Then I slip a piece of stiff wire bent at a 90-degree angle into the hole *as shown* and rotate the wire a full 360 degrees to see whether it hits anything. If it does, you can recheck the top plate and sole plate to try and identify what it is. If it doesn't, you can proceed to Step 3.

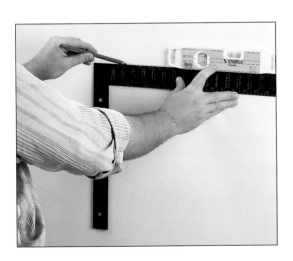

3 **Lay out opening** Mark the location of the wall studs, and then use a framing square and a torpedo level to mark the top and bottom of the opening *as shown*. If you're planning on a storage unit where the width is greater than 16", you'll need to remove the wall covering to expose the nearest studs so that you install the appropriate header to handle the weight that the studs were previously supporting.

4 **Cut wall opening** With the opening marked, now you can cut through the wall covering. The best tool for this job is a reciprocating saw fitted with a demolition blade. Cut on the waste side of the lines you drew in Step 3, and be careful as you near the wall studs. Cut the sides first, then the bottom, followed by the top. Carefully lift out the wall covering and set it aside for disposal. If the wall is lath and plaster, see *pages 84–85* for directions on how to remove it.

5 **Frame wall opening** For storage units that fit between a single pair of studs, measure and cut a header and a sill plate to span the studs. Tap each in place, and secure them to the wall studs with screws *as shown*. For wider openings, it's best to fit a header and sill plate between the studs and then add new cripple studs to frame the opening, as shown in framing the opening for the built-in medicine cabinet shown on *page 124*. If the wall you're working on is a load-bearing wall and you're removing more than one stud, make sure to support the ceiling joists with a temporary support (*see pages 86–87*).

6 **Build case** Now that the opening is framed, you can build the storage unit. This can be a simple box that's screwed together. I like to use melamine (particleboard with a thin plastic coating) for built-ins because it requires no finish. Cut two sides and a top and a bottom to fit the opening (*see the sample built-in on page 111 for the basic box*). Then attach the sides to the top and bottom with screws. **Note:** Special screws for particleboard are available at some hardware stores; these screws have deeper threads to cut into the particleboard and afford a better grip.

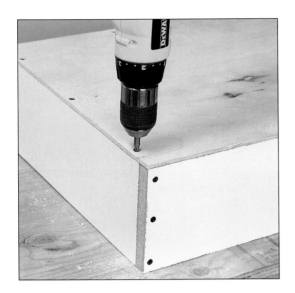

7 **Add back** To complete the storage unit, measure and cut a back to fit. Make sure the case is square by measuring diagonally across the back, then checking to make sure the opposite diagonal measurement is the same. If the two diagonals are not of equal length, rack the case until they are. Then attach the back to the case with glue and screws (or brads). Note: If you can't locate ¼"-thick melamine, you can apply a couple coats of white paint to one side of a piece of ¼" plywood (*shown here*) before attaching it to the case.

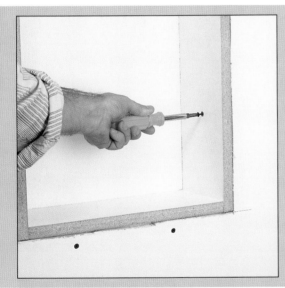

8 **Install unit** With the case built, the next step is to install it in the rough opening. (**Note:** See the sidebar *below* for instructions on drilling holes for shelf pins before installing the case.) If necessary, add shims between the case and the framing members to make it plumb and level. Adjust the case so that its front edges are flush with the wall covering. Then drill pilot holes through the case and into the framing members. Secure the case to the studs with screws *as shown*.

ADDING ADJUSTABLE SHELVES: PEGBOARD TIP

Regardless of the shape or size of your built-in storage unit, you'll want to add shelves. Instead of adding fixed shelves, I always recommend adjustable shelves. The only challenge with this is drilling accurately spaced holes for the shelf pins. Here's a foolproof way to do this. Cut a scrap piece of pegboard to fit roughly inside the case *as shown.* Then make marks on the pegboard to indicate where you'd like holes for shelf pins (mark these on both sides, as you'll need to flip the pegboard when drilling into the opposite side). Butt the pegboard up against the inside corner of the case, and drill holes where indicated; flip the pegboard over and drill the other side—the holes will be perfectly aligned.

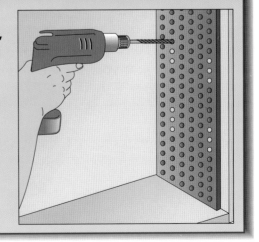

9 **Add trim** To cover the exposed edges of the case, hide gaps in the wall, and provide a decorative touch, I installed trim around the case. Measure and miter-cut one piece at a time. Attach each trim piece with finish nails or brads (I used an air-powered nailer), and then move on to the next piece. When all the trim pieces are attached, countersink the nails or brads with a nail set and putty over the holes with spackling compound.

10 **Install shelves** The last thing to do is to add the shelves. These can be melamine, hardwood, or glass (*like those shown here*). Measure between the sides of the case, subtract ⅛" for clearance, and then cut the shelves to length. I like to make the shelves about ⅜" narrower than the width of the case to provide a nice setback. Install shelf pins at the desired locations and set the shelves in place. **Note:** A nifty source for glass shelves are replacement sections for jalousie windows. The edges are already rounded over, and they're easy to cut to length.

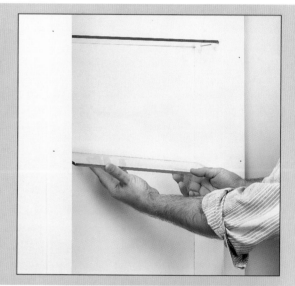

INSTALLING LIGHTING

Most built-in storage units will benefit from some form of recessed lighting. The lighting *shown here* is easy to install and will really brighten up any wall unit. Most lighting kits give you the option of surface-mounting the lights or recessing them like the pair of lights *in the photo at right.* Lighting like this works especially well with glass shelves, which allow the light to filter all the way down to the bottom of the case. Halogen lighting kits like the one used here are available at most hardware stores and home centers.

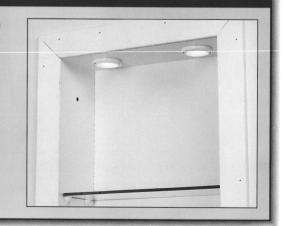

Built-In Medicine Cabinet

1 **Locate wiring/plumbing** The first step in planning to install a built-in medicine cabinet is to identify and locate the inevitable plumbing and electrical lines that can have an impact on the job. The bathroom *shown in the photo here* has an existing light above the sink that will have to be removed. The wiring inside the wall will most likely have to be moved as well. The new medicine cabinet we're installing has lights built in, so we can hook these up to the existing wiring.

2 **Mark opening** The next step is to locate the wall studs with an electronic stud finder and then mark the appropriate-sized opening for the new cabinet. A framing square and a torpedo level work well for this. **Note:** Although you can purchase a medicine cabinet to fit between a single pair of wall studs, these tend to be very narrow. The cabinet being installed here is 26" wide. This means that we will need to cut one of the wall studs and install some new framing.

3 **Cut opening** Once the opening is marked, you can cut through the wall covering. This can be done with a reciprocating saw, a circular saw, or a utility knife (*as shown here*). Although using a utility knife to cut through drywall is tedious, it creates the cleanest edge and disturbs the rest of the wall the least. **Note:** If the new medicine cabinet you're installing is wider than 16", cut the wall covering back far enough that the nearest wall studs will be exposed. This way you'll be able to add a new header and sill plate.

4 **Knock out wall** After the wall covering is cut, you can remove it with any of the demolition techniques described on *pages 82–85.* Make sure to turn off power to the room and remove any old electrical fixtures before beginning demolition work. Also, you may find that it's simpler in the long run to remove the wall covering all the way up to the ceiling—it's a lot easier to install new drywall than to try to patch a smooth transition between the new and existing walls.

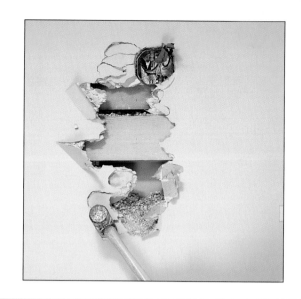

5 **Frame opening** To frame the new opening, start by measuring the distance between the wall studs and cut a header and sill plate from 2×4 stock. Cut away the interfering stud and insert the header and sill plate. (If you're removing more than one stud in a load-bearing wall, make sure to use temporary supports; *see pages 86–87.*) Secure the header and sill plate to the wall studs with screws. Then measure the distance from the header and sill plate, and cut one or two cripple studs to create a rough opening. Secure the cripple studs with screws *as shown.*

6 **Attach drywall** With the framing complete, you can install drywall to cover the exposed studs. Cut the drywall so the edges are flush with the inside edges of the rough opening, and secure them to the framing members with drywall screws. You may or may not wish to tape and finish the drywall at this point—it's a matter of personal preference. For step-by-step directions on how to finish drywall, *see pages 108–109.*

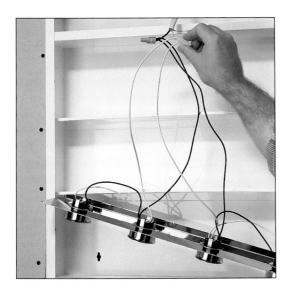

7 **Hook up electrical** Before you can install the unit, you'll need to make any electrical connections if the cabinet has built-in lighting *like the one shown here.* Follow the manufacturer's directions to the letter, and make sure to use an appropriate cable clamp to protect the wiring as it enters the metal case of the cabinet. Also, be sure to connect the ground wire to the ground lug on the cabinet to help prevent electrical shocks.

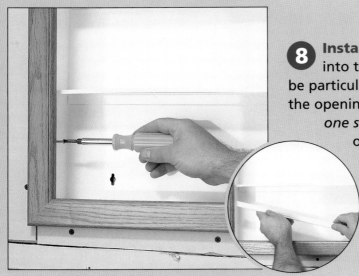

8 **Install unit** At this point you can install the cabinet into the opening. If your cabinet has built-in lighting, be particularly careful to feed the electrical wiring into the opening as you install the unit. Some cabinets (*like the one shown here*) are designed for surface mounting or recessed mounting. Often with this style cabinet, you'll need to drill mounting holes in the side of the cabinet for the screws that attach the cabinet to the framing members. After you've secured it with screws, install any interior shelves (*inset*).

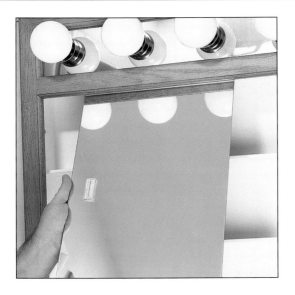

9 **Add doors** The final step in completing the installation is to add the doors. These are typically mirrors (*as shown here*) or glass. In either case, you'll usually have to first attach handles to each door with double-faced tape provided by the manufacturer. Then you'll most likely need to install the tracks that the doors ride in. They're usually held in place with double-sided tape as well. Push one of the doors up into the track, pivot it in, and set it in the lower track. Repeat for the remaining door.

Glossary

Air nailer – any air-powered tool that shoots fasteners into a workpiece. When activated, compressed air forces a piston with an attached driver blade to drive the next fastener.

Balloon framing – in a balloon-framed structure, the wall studs run unbroken from sill to top plate, no matter how many stories the structure has.

Beam – a steel or wood member that's installed horizontally to support part of a structure's load.

Bird's mouth – a notch cut near the end of a rafter to fit over a cap plate.

Blocking – horizontal blocks that are inserted between studs every 10 vertical feet to prevent the spread of fire in a home.

Bridging – steel braces or wood blocks that are installed in an X pattern between floor joists to stabilize the joists.

Casing – boards that line the inside of a doorway, or the trim around the door opening.

Chords – framing members that make up the two sides of the roof and the base of a triangular truss.

Collar tie – a horizontal framing member installed between rafters to add stiffness.

Column – a metal or wood vertical member designed to support part of a structure's load.

Cornice – the part of the roof that overhangs a wall; also called the roof overhang.

Cripple studs – short vertical studs installed between a header and a top plate or between the bottom of a rough sill and the sole plate.

Dormer – a shedlike structure that projects out from a roof to provide additional attic space.

Double top plate – a double layer of 2-by material running horizontally on top of, and nailed to, the wall studs.

Drip edge – a bent metal strip that fits over the edge of the roof to direct rain away from the roof edge and underlying walls.

Eaves – the part of the roof that projects past the supporting walls.

Fascia – a trim piece fastened to the ends of the rafters to form part of the cornice.

Flashing – thin metal used to bridge gaps between the roof and framing or shingles and framing; also used to line valleys to shed water.

Footing – typically a poured concrete base on which the foundation of a structure is built.

Frieze board – trim pieces installed directly beneath the rafters to provide a nailing surface for the soffits and corner trim.

Girder – a horizontal steel or wood member used to support part of a structure's load.

Header – a horizontal framing member that runs above rough openings to take on the load that would have been carried by the wall studs; may be solid wood, be built up from 2-by material, or be an engineered beam such as MicroLam or GlueLam.

Hip rafter – any rafter that runs at a 45-degree angle from the end of the ridge to a corner of the structure.

Jack rafter – short rafters that run between two rafters or a rafter and a top plate.

Jack stud – a stud that runs between the sole plate and the bottom of the header; also referred to as a trimmer stud.

Joist – framing lumber that's installed horizontally on edge to which subfloors are attached.

Joist hanger – often referred to as metal framing connectors, these are designed for use on 2-by projects where you need to quickly attach framing members together. Connector manufacturers offer an unbelievable assortment of anchors and ties for almost every conceivable application.

King stud – the wall stud to which the jack stud is attached to create a rough opening for a window or door.

Load-bearing wall – a load-bearing wall helps support the weight of a house; all of the exterior walls that run perpendicular to the floor and ceiling joists in a structure are load-bearing walls, and any interior wall that's located directly above a girder or interior foundation wall is load-bearing.

Non-load-bearing wall – a non-load-bearing wall does not help support the weight of the structure; also referred to as partition walls, they have relaxed design parameters and code requirements, such as wider stud spacing (24" vs. 16" on center) and smaller headers.

Overhang – the end of the rafter that projects beyond the building line; typically enclosed by a soffit.

Pier – a round or square concrete base used to support columns, posts, girders, or joists.

Pitch – the rise of the roof over its span.

Platform framing – a platform-framed structure is built one story at a time; each story is built upon a platform that consists of joists and a subfloor.

Post-and-beam framing – Post-and-beam construction is easily identified by its use of large, widely spaced load-carrying timbers; also referred to as post-and-girt or post-and-lintel.

Powder-activated nailer – often referred to as a power hammer or a powder-actuated nailer; it's basically a gun barrel that's activated by a trigger or by striking it with a hammer. In either case, this discharges a hardened steel fastener expelled by a bulletlike powder load.

Rafter – a framing member that connects to the ridge board and extends down to the double top plates of the walls; rafters typically tie into the ceiling joists, which prevent the walls from bowing out under the load of the roof.

Ridgeboard – the horizontal board the defines the roof's highest point or ridge.

Rim joists – the joists that define the outside edges of a platform, typically nailed to the ends of floor joists.

Ring-shank nails – nails that are manufactured with rings along the shank to provide extra grip.

Rise – the vertical distance between the supporting wall's cap plate and the point where a line, drawn through the outside edge of the cap plate and parallel to the roof's edge, intersects the centerline of the ridge board.

Rough opening – an opening that's sized to accept a window or door; a horizontal framing member called a header is installed to assume the load of the wall studs that were removed. The header is supported by jack studs that are attached to full-length wall studs.

Rough sill – a horizontal framing member that defines the bottom of a window's rough opening.

Scribing – a layout technique used to copy the imperfections of a wall to flooring so the flooring can be cut to butt snugly against the wall.

Sheathing – panel material, typically plywood, that's applied to the exterior of a wall prior to the installation of siding.

Shim – a thin piece of wood that, when driven behind a surface, forces it to become level or plumb.

Slope – the rise of the roof over its run, expressed as the number of inches of rise per unit of run (typically 12"); 8 in 12 means a roof rises 8" for every 12" of run.

Soffit – the board that runs the length of a wall, spanning between the wall and the fascia on the underside of the rafters.

Sole plate – a horizontal 2-by framing member that is attached directly to the masonry foundation or flooring; also referred to as a sill plate or mudsill.

Stud – a vertical 2-by framing member that extends from the bottom plate to the top plate in a stud wall.

Subfloor – the first layer of a floor structure fastened directly to the joists or to a concrete slab.

Temporary supports – temporary supports bear the weight a wall normally would until a new support system can be installed (such as a new header or beam).

Top plate – a horizontal 2-by framing member that's nailed to the tops of the wall studs.

Underlayment – a smooth surface laid on top of the subfloor to accept flooring; can be sheets of plywood, foam or cork, or cement board.

Vapor barrier – plastic sheeting installed between walls to prevent moisture from entering and damaging the structure.

Index